STUDIES IN ORIENTAL RELIGIONS

Edited by Walther Heissig and Hans-Joachim Klimkeit

Volume 10

RAGNHILD BJERRE FINNESTAD

IMAGE OF THE WORLD AND SYMBOL OF THE CREATOR

On the Cosmological and Iconological Values of the Temple of Edfu

1985

OTTO HARRASSOWITZ · WIESBADEN

RAGNHILD BJERRE FINNESTAD

IMAGE OF THE WORLD AND SYMBOL OF THE CREATOR

On the Cosmological and Iconological Values of the Temple of Edfu

1985

OTTO HARRASSOWITZ · WIESBADEN

The series STUDIES IN ORIENTAL RELIGIONS is supported by

Institute of Comparative Religion
Bonn University

Institute of Central Asian Studies
Bonn University

in collaboration with

Institute for Advanced Studies
of World Religions New York

Institute of History of Religion
Uppsala University

Donner Institute, Academy of Åbo
Åbo, Finland

Institute of Oriental Religions
Sophia University, Tokyo

Department of Religion
University of Hawaii

CIP-Kurztitelaufnahme der Deutschen Bibliothek

Finnestad, Ragnhild Bjerre:
Image of the world and symbol of the creator :
on the cosmolog. and iconolog. values of the
temple of Edfu / Ragnhild Bjerre Finnestad. –
Wiesbaden : Harrassowitz, 1985.
(Studies in oriential religions ; Vol. 10)
ISBN 3-447-02504-2
NE: GT

Gedruckt mit Unterstützung des Norwegian Research Council
(Norges almenvitenskapelige forskningsgråd)
Satzherstellung: PM-team, 5342 Rheinbreitbach

Printed in Germany
Sigel: StOR

Otto Harrassowitz GmbH & Co. KG
Kreuzberger Ring 7c-d, D-65205 Wiesbaden
produktsicherheit.verlag@harrassowitz.de

Contents

ABBREVIATIONS IX

ACKNOWLEDGEMENTS 1

INTRODUCTION 3

1. The metaphorical and symbolical meanings of the Egyptian temple 3
2. Source material 4

A. THE TEMPLE IS A COSMOPHANY 7

I. *The temple represents the habitation of god* 7

II. *The temple represents cosmos* 8

a) Outline of the cosmological imagery of the habitation of god 8
   1. The geography represented by the temple of Edfu . . . 8
      a. The temple is Egypt 8
      b. The temple is Edfu 10
   2. The topography represented by the temple of Edfu . . . 10
b) Explication of the cosmological imagery 16
   1. The source material evaluated 17
   2. The cosmogony myth of Edfu 24
      a. Translation 25
      b. Definition of the cosmos created 42
         1) The topography of the cosmos 42
         2) The geography of the cosmos 46
c) The creation of the dwelling of god 52
   1. As presented by the cosmogony myth 52
   2. The ritual foundation of the temple connotes the cosmogony 56
   3. The nomenclature of the temple connotes the cosmogony 64

III. *The ritual aspect of the cosmogony* 68
a) The mode of creation 68
b) The ritual bestowal of names on the temple 74

IV. *Conclusion*
The temple reveals the sacred nature of the world of man . . 78

B. THE TEMPLE IS A THEOPHANY 79

I. *The mythological background of the theophanic character of the temple* 79
a) The mythical theophany 79
b) The mythical theophany implies a theogony 89
1. The theogonic appearance 90
2. The theogonic invocation 92
c) Conclusion 93

II. *The temple-cultic theogony* 94
a) The temple as a stage for the theophany 94
b) The revelation of the statue: the god living in his dwelling . 96
1. The daily appearance 96
2. The theogonic nature of the daily appearance 99
3. The seasonal appearance 101
4. The latent life of the gods is located in the temple . . . 104
c) The revelation of the temple-building 111
1. The appearance of the cosmos-constituting gods 111
2. The theogonic invocation 113
d) The theogonic role of the king 120

III. *The iconological aspect of the temple* 121

C. THE TEMPLE IS A SYMBOL OF GOD 124

I. *The iconological character of the temple*. 125
a) On the definition of icon 125
b) Remarks on the hermeneutical concepts applied to the material in the analysis 126
c) The iconological functions of the temple 128
d) The temple undergoes the consecration ritual of the icon . . 131
e) The temple-cultic ka-life of the god 134
f) Personification of the ka-life of the temple 138

II. *Distinguishing features of the icon as regards cognitive content and cultic functions* 142
a) The temple is the icon of a creator immanent in his work of creation 142
b) The temple formulates a henotheistic concept of god . . . 145
c) The temple is the icon of a dynamic god 147
d) The temple formulates god as a cultic phenomenon . . . 148

Abbreviations

AÄG	= E. Edel, Altägyptische Grammatik. Analecta Orientalia 34/39, Roma 1955/1964.
AAWLM	= Abhandlungen der Akademie der Wissenschaften und der Literatur in Mainz, Wiesbaden.
AcOr	= Acta Orientalia, Copenhagen.
ADAIK	= Abhandlungen des Deutschen Archäologischen Instituts Kairo, Glückstadt, Hamburg, N. Y.
AfO	= Archiv für Orientforschung, Berlin; from vol. 15: Graz.
AHAW	= Abhandlungen der Heidelberger Akademie der Wissenschaften, Heidelberg. ,
AL	= D. Meeks, Année Lexicographique Égypte Ancienne, Paris.
ASAE	= Annales du Service des Antiquités de l'Égypte, Cairo.
BdE	= Bibliothèque d'Étude, Institut Français d'Archéologie Orientale, Cairo.
BIFAO	= Bulletin de l'Institut Français d'Archéologie Orientale, Cairo.
The Book of the Dead	= E. Naville, Das aegyptische Todtenbuch der 18.–20. Dynastie, Berlin 1886.
BSFE	= Bulletin de la Société Française d'Égyptologie, Paris.
CD	= R. O. Faulkner, A Concise Dictionary of Middle Egyptian, Oxford² 1964.
CdE	= Chronique d'Égypte, Brussels.
DAWW	= Denkschriften der Kaiserlichen Akademie der Wissenschaften in Wien, Phil.-hist. Kl., Wien.
De Iside	= Plutarch, De Iside et Osiride, edited and translated by J. Gwyn Griffiths, Cambridge 1970.
DG	= H. Brugsch, Dictionnaire géographique de l'ancienne Égypte, Leipzig 1879; H. Gauthier, Dictionnaire des Nomes Géographiques, 7 vols. Cairo 1925–1931.
E	= M. de Rochemonteix, Le Temple d'Edfou I, MMAF 10, 1897; E. Chassinat, M. de Rochemonteix, Le Temple d'Edfou II and III, MMAF 11 and 20, 1918 and 1928; E. Chassinat, Le Temple d'Edfou IV–XIV, MMAF 21–31, 1929–1934.
GM	= Göttinger Miszellen, Göttingen.
JEA	= Journal of Egyptian Archeology, London.
LÄ	= Lexikon der Ägyptologie, Wiesbaden.
MÄS	= Münchner Ägyptologischen Studien, Berlin.
MDAIK	= Mitteilungen des Deutschen Archäologischen Instituts, Abteilung Kairo; until 1944: Mitteilungen des Deutschen Instituts für Ägyptische Altertumskunde in Kairo, Berlin, Wiesbaden; from 1970: Mainz.

MIFAO	= Mémoirs publiés par les Membres de l'Institut Français d'Archéologie Orientale du Caire, Cairo.
MMAF	= Mémoirs publiés par les Membres de la Mission Archéologie Française au Caire, Paris.
MO	= E. A. E. Reymond, The Mythical Origin of the Egyptian Temple, Cambridge 1969.
MVÄG	= Mitteilungen der Vorderasiatisch (-Ägyptisch)en Gesellschaft, Leipzig and Berlin.
NAWG	= Nachrichten von der Akademie der Wissenschaften zu Göttingen, Göttingen.
ÖAW	= Österreichische Akademie der Wissenschaften, Wien.
OLA	= Orientalia Lovaniensia Analecta, Leuven.
ON	= Orientalia Neerlandica. A Volume of Oriental Studies published under the Auspices of the Netherland's Oriental Society on the Occasion of the 25th Anniversary of its Foundation, Leiden 1948.
PIFAO	= Publications de l'Institut Français d'Archéologie Orientale, Cairo.
Porter and Moss	= B. Porter and R. L. B. Moss, Topographical Bibliography of Ancient Egyptian Hieroglyphic Texts, Reliefs and Paintings VI, Oxford 1939.
PSBA	= Proceedings of the Society of Biblical Archeology, London.
Pyr. t.	= K. Sethe, Die altägyptischen Pyramidentexte, Leipzig 1908–1922.
RÄRG	= H. Bonnet, Reallexikon der ägyptischen Religionsgeschichte, Berlin 1953.
RAr	= Revue Archéologique, Paris.
Rec. trav.	= Recueil de Travaux Rélatifs à la Philologie et à l'Archéologie Égyptiennes et Assyriennes, Paris.
RHR	= Revue de l'Histoire de Religions, Paris.
StG	= Studium Generale, Berlin, Heidelberg, N. Y.
Stud Aeg, Rome	= Studia Aegyptiaca, Rome.
Stud Aeg, Budapest	= Studia Aegyptiaca, Budapest.
TÄB	= Tübinger Ägyptologische Beiträge, Bonn.
WZKM	= Wiener Zeitschrift für die Kunde des Morgenlandes, Wien.
ZÄS	= Zeitschrift für Ägyptische Sprache und Altertumskunde, Leipzig, Berlin.

Signs applied in translations

[ ]	Enclosing damaged words or parts of words.
⌜ ⌝	Enclosing translations which are open to question.
< >	Enclosing words omitted by the scribe.
( )	Enclosing words added to clarify the sense.

Acknowledgements

This book is a slightly revised version of my doctoral thesis of 1982, defended in 1983.

The major part of the research work was carried out at the Institute of Egyptology in Copenhagen, and I want to give my sincere thanks to the members of this institute for their hospitality and great helpfulness in granting me every opportunity to avail myself of all practical facilities as well as their professional expertise. In particular, I want to acknowledge my great debt to Professor Jürgen Osing, for his generous assistance with the translation of the fundamental text under consideration in my monograph. His contribution is invaluable, and I doubt if there is any way I can adequately thank him. Errors of interpretation are strictly my own.

Also, I wish to express my warm thanks to Universitetslektor Paul John Frandsen, for his useful advice on grammatical problems and on questions concerning my translation.

My thanks are due to Anne Zeeberg, the librarian at the institute, who was always ready to help, and to Torben Holm Rasmussen, whose knowledge came to my aid more than once.

I want to thank Professor Jan Bergman, Uppsala, who was my First Opponent, for his useful comments and suggestions.

Further, I thank Erik Iversen, Copenhagen, for inspiring discussions on late Egyptian religion, and my colleague Lisbeth Mikaelsson, Bergen, who, although she did not read my manuscript, gave her support through conversations that made it possible for me to clarify important issues in this monograph.

I extend my deepest gratitude to Professor Richard Holton Pierce, Bergen, who read through my manuscript and never hesitated to offer his observations and advice on both substance and formal matters. His encouragement throughout my work has been an important support.

I thank l'Institut Français d'Archéologie Orientale in Cairo for its obliging assistance in procuring copies of illustrations from Chassinat's editions of the Edfu texts.

Finally, I want to express my thanks to the Norwegian Research Council for Science and the Humanities which supported my work with a research grant and which has contributed to the publication of this book.

Ragnhild Bjerre Finnestad

Introduction

1. The metaphorical and symbolical meanings of the Egyptian temple

The concepts pertaining to the Egyptian temple are manifold and woven together into a complex web which at first sight is not easy to analyse. Yet two basic metaphorical meanings can be discerned straight-away as the warp and the weft; the one being that the temple is a habitation of god, and the other being that the temple is a *cosmos.** When appearing as a habitation of god, the temple is called by designations such as the House of God; the idea is also expressed by its architecture, as it is fashioned analogous to the dwellings of man. When appearing as cosmos, the temple is called by names identifying it with a cosmological geography and topography, and it has frequently been pointed out that it displays architectural features identifying it with the landscape of cosmos: its roof is the sky, its floor is the soil of Egypt from which pillars "grow" like vegetation. Parallels have been drawn with the cosmos imagery of temples in other cultures.

This concurrence of the metaphorical values of divine dwelling and of cosmos deserves attention, as it expresses a central theologumenon in Egyptian religion: here we have a cult-topographical phrasing of the immanence of the Egyptian gods which carries a far wider significance than does the formally similar double imagery of the Christian church – which is also a House of God and an image of man's world. The Egyptian temple belongs within a monistic ontology, where divine life and cosmic life coincide and where there is no ontological basis for god outside cosmos. The *iconological* functions of the temple should be understood within this frame of reference: As will be demonstrated, the temple also has symbolical meaning: it can appear as an image of the god immanent in cosmos – be conceived of as the concrete form of the god of cosmos.

The aim of this monograph is to examine the metaphorical values of the temple as well as its functions as a symbol of god. The analysis will be done with regard to an explication of the concept of god thus presented in the temple ideology. The temple of Horus at Edfu is chosen as the object for this examination.

* In this monograph *cosmos* is used as a technical term denoting the place of man's habitation: "man's world". It is an analytical structure; the Egyptians themselves had no word for cosmos.

2. Source material

a) The kind of material utilized in the analysis

Theological expositions are the sources traditionally preferred when studying the concepts of god. However, this kind of source material does not give satisfactory information regarding many religions, even when they have written traditions, because it may not be a representative mode of expression for those religions. This is the case with the religion of ancient Egypt. The ancient Egyptians were not devoted to theology in the strict sense of the word: i. e. an intellectual discipline for the rational, theoretical explication of god. Therefore, this kind of source material is rather meagre, if existant at all. Generally speaking, the inclination of Egyptian religion is towards its cultic performance, not discursive thought. The bulk of textual information concerning the Egyptian gods therefore comes from mythological and ritual texts belonging within a cultic context.

For the same reason, an equally important source of information as regards the Egyptian concepts of god is the non-textual material which does not formulate the religious beliefs in words, but visualizes them through concrete forms. Thus the temple building with its topographical layout, architecture, and decoration is a highly relevant source of information on the Egyptian gods; since it is the main stage of the cultic performance.

This non-textual material cannot, however, be interpreted isolated from texts referring in some way to its religious meaning, just as the texts cannot be evaluated adequately without reference to the concrete cultic material. There is an interdependency between the different types of material. Therefore, the textual and non-textual materials utilized in this analysis are approached with the purpose of discerning their capacity to constitute a whole.

The study is limited to aspects of the temple which are of special relevance to the theme of divine immanence, and this implies a selection of texts and topographical, architectural and decorative features. The textual material occupies a central position. As a hermeneutical consequence of the inclusive approach to the sources, these are interpreted in their context as an integral part of the temple-cultic whole. The aim is that this approach will contribute to a better understanding of the texts.

b) Reasons for choosing the Edfu temple as an object for analysis

For the purpose outlined above, it is necessary to have access to an intact temple; and the choice of Edfu is thus a natural one. Most Egyptian temples are variously in ruins. The Ptolemaic temples are an exception: some of them are very well preserved, especially the temple of Edfu which offers a unique opportunity of studying an Egyptian temple in its entirety. Also, since it was built in a relatively short period of time, it possesses a singular unity and coherence – in contrast to temples built over many centuries, and which had been continuously added to and altered. The sacral-architectonic intention of the building can be apprehended, and from it invaluable information pertaining to the religious ideology behind the temple can be obtained.

The Edfu temple was completed within the span of 180 years. Construction started under Ptolemy III in 237 B. C. By 142 the sanctuary and the adjoining rooms, the hall of offerings and the hypostyle hall, were raised. By 124 the pronaos, the surrounding wall, and the forecourt (plate 5) were built. Finally, in 116, the great pylon was erected (plates 3 and 4). The inauguration took place in 71, but the temple was not fully completed with reliefs and embellishments until 57 B. C. Inasmuch as the temple was concluded in that year, it belongs to a late period of Egyptian religion. This does not mean that it represents a temple ideology alien to "authentic" or "ancient" Egyptian religion. On the contrary, the ideas underlying its architecture, embellishments and texts are typically and genuinely Egyptian[1]. Notwithstanding architectural features characteristic of the period[2], it is apparent that the temple was built according to a model that can be recognized in the New Kingdom temples. It displays a traditional sacral style of building.

On the other hand, as has been pointed out by many[3], the Edfu temple represents a further development of the earlier tradition, a fact which in itself is a guarantee that tradition was unbroken. The people who built this temple did not mechanically copy what had become obsolete and incomprehensible to them; they enlivened the traditions with an inspiration derived from religious understanding. This renders the temple no minor source of information

1 Already pointed out by de Rougé, who characterized the temple as "un vrai modèle de l'architecture religieuse du peuple égyptien", 1865. Cf. also Noshy 1937 pp. 70, 80; Blackman and Fairman 1941 p. 426 n. 127; Baldwin Smith 1968 p. 181. – One of the reasons for this continuous tradition apparently is to be found in the cultic purpose of the temple, for as Daumas has pointed out: "Mais si les modes changent, les rites religieux sont plus stable", 1981 p. 261.

2 A Ptolemaic feature is e. g. the pronaos: a columnar vestibule in front of the hypostyle hall, larger and loftier than the latter; the façade of this frontal vestibule was a screen wall with pillars, E IX pl. X. See Baldwin Smith 1968 pp. 179, 183.

3 For example, Winter 1968 and 1969. Winter has shown that the texts and decorations of the temple walls are in accordance with a theological programme set up in Ptolemaic times as a development of older traditions: he concludes from his studies that "der Ägypter auch in dieser späten Zeit das ihm zur Verfügung stehende theologische Material souverän beherrscht hat" (1969 p. 125). Cf. also E. Otto 1969 p. 120; Sauneron and Stierlin 1978.

as regards Egyptian religion. The temple of Edfu presents the basic Egyptian religious ideas with artistic precision and consequence – the trade mark of a conscious intention and a most welcome formal characteristic from the point of view of the interpreter.

Another reason for choosing the Edfu temple as an object for analysis is the fact that its texts comprise cosmological records of rare and outstanding value for our study, an advantage probably due to the high degree of preservation of the temple. In the texts of Edfu, the relationship between the dwelling of god and the cosmos of man can be explicitly traced. Especially the cosmogony texts are of importance in this respect. Any study of the immanence of the Egyptian gods will uncover a rewarding source of information here.

These texts are interpreted on the basis of their explicit content. The question whether they may have been explained by the priests in a way that would accommodate Hellenistic ideas will not be discussed, partly because this question belongs to another kind of study, and also because it cannot be answered by our material. As is the case with the form and decoration of this temple, the texts have developed within the Egyptian tradition. Though the language is not classical Egyptian, it can be said to be in agreement with the fundamental canons of the language, and the conceptual quality of the texts is unmistakably Egyptian. In addition, a Hellenistic re-interpretation of the texts would hardly seem possible, since these texts form part of a temple-cultic whole that concretizes the monistic ontology characteristic of Egyptian religion, as this monograph will try to show.

A. The temple is a cosmophany

I. The temple represents the habitation of god

The dwelling imagery of the Edfu temple is apparent. Its architectural features conform with those of the living quarters of man; the activities taking place here are similar to those taking place in the house of man. It is built like a manor – with a courtyard, pillared halls, and inner apartments[4]. It contains a dining-room (offering hall), library, cloakrooms, and store-rooms. There are private apartments for the owner and the members of his family and for the rest of the household[5]. In this house god dwells, and the temple rituals represent his life: in the inner room he sleeps during the night[6], he is awakened in the morning, undergoes his toilet and has his breakfast[7]. The temple is *The Mansion of God (ḥt-nṯr)*[8]: a private home for him and his family, not for the public congregation.

Today, the temple of Edfu is deserted and lifeless, a grey building of stone. When it was inhabited, it was filled with the noises and smells of a big household; the building was brightly painted, and like all large manors it was surrounded by vegetation: it had trees, bushes and flowers, and a pond[9].

Also with regard to its functions, the conformity with a manor was apparent in those days: the temple possessed extensive tracts of land – fields and meadows yielding crops for the benefit of the people working in this household, and for its big herds of cattle[10]. The temple was an estate conducting all kinds of activities, the centre of a sizeable landed property[11]. It played an important role in food storage and redistribution in the Egyptian society[12]. Even industrial enterprises were undertaken, such as brick-manufacturing,

4 Steindorff 1896; Badawy 1968.

5 There is a chapel for Hathor, his consort (E IV 5,7–8; VII 14,3), a chapel for Re (E IV 5,8), chapels for Osiris (E IV 5,5; 13,11; VII 13,3), a chapel for Min (E IV 6,2; VII 15,9): there is a hall for the Ennead (E IV 13,13).

6 E. g. E III 87, 12f: "... he sleeps in Behdet every day". Cf. also the reference given by Blackman and Fairman 1941 p. 426 n. 127.

7 Alliot 1949. Further, Moret 1902.

8 E IV 95,9; E VI 168,10; VII 200,15; RÄRG p. 778.

9 Gothein 1966; Gallery 1978.

10 Roeder 1960.

11 W. Otto 1905.

12 Butzer 1978 p. 16. Cf. also Posener-Krieger, and Janssen, in Lipinski 1979.

baking, brewing, and weaving of textiles[13]. From this it appears that the dwelling of god was an important cultural and economic institution, functioning in the combined rural and urban enterprise typical of Egyptian culture. This habitation of god can be regarded as an epitome of Egyptian civilization.

II. The temple represents cosmos

a) Outline of the cosmological imagery of the habitation of god

The dwelling of god represents the world of man, and can thus be said to portray both the place where god lives and the place where man lives.

The cosmological aspect of the Egyptian temple is generally recognized. As early as 1894 Rochemonteix drew attention to it in his inspired "Le temple égyptien"[14]. Topographical allusions of the architecture and decoration have been focused upon when the cosmological meaning of the temple is to be demonstrated, although, as shall be shown, the landscape presented by the architecture and decoration is not strictly speaking "cosmic". In the nomenclature of the temple, however, a cosmological geography is undoubtedly reflected, for it is given names to identify it with Egypt and the town to which it belongs. Also, certain decorative elements give the temple precise geographical identity.

1. The geography represented by the temple of Edfu

The temple bears names and epithets that identify it with the land of Egypt and the town of Edfu.

a. The temple is Egypt

One of the well-known names of Egypt included in the nomenclature of the temple is *The Two Lands (t3wj)*[15], a name which reflects the notion that Egypt consists of two parts, The North and The South. This notion has a central

13 W. Otto 1905, I, ch. 4.

14 De Rochemonteix 1894 pp. 1–38. The first one to note the cosmological meaning of the temple was possibly Maspero 1887, p. 98 ff.

15 E. g. E I 13, right, below. Wb V 217.

position in the religious ideology surrounding Egyptian culture[16], and the dual-and-united land is the geographical frame of reference of many religious events, especially those connected with kingship, as the king was the "unifier" of the complementary opposites of The North and The South[17]. The concept of the united lands is important in the temple ideology as well, and numerous decorative traits reflect this concept. Two particularly conspicuous traits are the plant emblems and nome-figures belonging to The North and The South, engraved on the base of the temple. The plant emblems of the papyrus, representing the North, and of the lily, representing the South, run all around the base, on the outside[18] as well as the inside[19], alternately, as if outlining the border of the united land. The nomes of Egypt, depicted as being personified, march around the base[20], arranged according to their geographical distribution; thus the nomes of the North appear on the western-and-northern half of the temple, while those of the South appear on the eastern-and-southern half[21], an arrangement which directs attention to the dual character of the land.

Another temple name with a geographical implication is *The Horizon (3ḫt)*. Although the designation indicates a complex set of ideas[22], the geographical reference is clear: the name connotes the utmost limits of the country. This is rendered explicit in the designation *The Horizon of The Two Lands (3ḫt t3wj)*[23]. *The Horizon* can also be applied to parts of the temple, which from an architectural point of view can be said to correspond metaphorically to this cosmological topos. Thus, the surrounding wall of the temple can be called *The Horizon* since it is the utmost limit of the temple proper; with this name the wall is likened to the horizon of Egypt, the utmost limit of the geographical cosmos. As this is also the place where the sun rises to unite with the heaven of The Two Lands, the name *The Horizon* has the additional connotation of a place of transition from chaos to cosmos. Within the geographical context, however, this meaning is often synonymous with confrontation, since the foreign

16 Cf. E. Otto 1938.

17 Frankfort 1948; Schäfer 1943.

18 Sauneron and Stierlin 1978 p. 28–29.

19 In the hypostyle: E II pls. XL a–h; in the central vestibule: E I pl. XXXI; in the sanctuary: E I pls. XI–XII; Sauneron and Stierlin, *op. cit.* p. 119.

20 The surrounding wall, inside: E X pls. CLV, CLVI, CLVIII; the courtyard: E X pls. CXXIII–CXXV, CXXVIII–CXXX; the pronaos, exterior: E X pls. CIX–CX; the naos, exterior: E X pls. XCIV–XCVI, XCIC–CI; the corridor around the *St-wrt*-sanctuary: E I pls. XV–XVII.

21 De Rougé 1865–1874; Montet 1957 p. 14 ff.; Beinlich 1976.

22 The concept of *3ḫt* comprises a wider area than the geographical one. For additional meanings, see Gutbub 1973 p. 299 n. b.

23 E. g. E II 34,2.

countries, inimical to The Two Lands, are located on the other side of the horizon. Thus, when the pylon is seen as The Horizon[24] in the geographical sense, i. e. the gateway to Egypt, the confrontation aspect is especially prominent, being visually displayed in the reliefs on the pylon, the most conspicuous of which depict Pharaoh slaying the foreign foes as they try to invade Egypt[25]. Thus the temple gate has the double meaning of confrontation between Egypt and the outside world, and of transition from the outside world into Egypt.

Other designations of Egypt are also applied to the temple, but the names that have been commented on are the ones recurring in the texts studied.

b. The temple is Edfu

The noticeable geographical equation in the textual material is with the town of Edfu and its surroundings[26]. The temple and the town share several names, the best known being *Ḏb3*[27], *Bḥdt*[28], *Wṯst*[29] and *Wṯst-Ḥr*[30].

The name of *Ḏb3* also denotes the nome of which Edfu is the capital city, and the temple name *Ḏb3* can have this reference too[31]. *Wṯs-Ḥr* is, however, the mose frequently used designation of the nome[32].

According to the above noted geographical definitions of cosmos, "man's world" is identical with the world of Egyptian man, as a native of The Two Lands and an inhabitant of Edfu.

2. The topography represented by the temple of Edfu

In architecture and decoration the temple depicts a landscape. As noted above, this was already noticed by Rochemonteix who attributed a cosmological

24 RÄRG p. 787; LÄ III 4. The two towers of the pylon can represent the mountains of the east and the west, Derchain 1966 p. 18.

25 E IX pl. III; E XIV pls. DCLXVI, DCLXVII. – There are impressive examples of the motif on the walls of the temples from the New Kingdom. See Kuentz 1928–1934; Wildung, LÄ II 146 f.

26 Also this cosmological meaning is in accord with ancient Egyptian temple ideology: each temple represented the cosmos of its geographical location; cf. Frankfort 1969[6] p. 152 ff.

27 Brugsch DG p. 921 f., 1349 f.; Gauthier DG VI p. 126 f. Wb V 562. In Coptic: θΒω (B), ΤΒω (S), W. Westendorf, *Koptisches Handwörterbuch*, Heidelberg 1965–1977.

28 Brugsch DG p. 978, 1267 f.; Gauthier DG II p. 27; Wb I 470.

29 Brugsch DG p. 964; Gauthier DG I p. 210; Wb I 384,9.

30 Gauthier DG I p. 210.

31 E VI 112,6.

32 Gauthier DG I p. 210; Montet 1961 p. 30 f.; Wb I 384,10.

significance to the temple-landscape; this interpretation is, indeed, near at hand. The extensive use of plant forms in construction and embellishment is strikingly suggestive of vegetation. The papyrus and lilies in the decoration on the base evoke the impression of plants growing out of the site. More prominent, though, are the columns of the temple. They are formed in the shapes of papyrus-plants, lilies, and reeds, and are clustered together in a way most impressively indicating vegetation: they border the forecourt, as if bordering a pond[33] (plate 4); and they practically fill the space within the two halls[34] (plate 6). The metaphorical meaning of the columns cannot be overlooked. It has been pointed out, moreover, that their supporting function is camouflaged by the abacus, whereby the impression is created that the columns do not reach the roof[35] – to the advantage of their representational meaning of being vegetation growing towards the sky.

However, an important fact should be considered. The landscape presented is not that of cosmos, i. e. the cultivated earth of Egypt, the soil upon which and from the produce of which Egyptian man lives, but rather that of *chaos**, i. e. the inundated and still uncultivated soil. Although names and certain decorative traits identify this building geographically with cosmos, its architecture and the greater part of its decoration identify it topographically with chaos.

Because of this chaos-aspect, the statement that the temple represents cosmos should be modified. The temple-building as such is *cosmos in the state of chaos*. This ambivalent compound denotes accurately the metaphorical meaning of the temple-landscape, a meaning which has not, strangely enough, received the acknowledgement it deserves, although it is of essential importance in our understanding of the functioning of the temple.

The chaos aspect of the temple is easily recognized. Its interior conforms to chaos in its two main characteristics according to Egyptian cosmology: it is a place of water and of darkness[36]. It is an *aquatic* landscape. As an element of chaos, water is prior to the cosmic element of earth; to be more precise, the inundation water of the Nile is the element from which the earth emerges fertilized. The floor of the hypostyle halls is not bare earth, but inundated earth; its vegetation is a marsh of reed and papyrus[37], not the life-

* *Chaos* is in this monograph used to denote the ontological stage preceding cosmos.

33 E IX pls. V, VI.

34 E IX pl. VIII.

35 First pointed out by Borchardt 1897.

36 Grapow 1931; Brunner 1955a; Hornung 1956; Brunner 1957.

37 To enhance the effect of vegetation growing out of water, a water line painted in blue could be indicated on the base of the columns, see Borchardt 1902–1903; Täckholm and Drar 1950 p. 111.

sustaining crops of the tilled earth. Similarly, the floor of the forecourt does not represent earth but a body of water; moreover, on the base of the walls of the court sacred boats are depicted[38]. – Further indications of aquatic imagery are the nome-representations walking around the temple on its base: they are in the shape of Nile-divinities[39]; the nomes of Egypt are represented here as inundated area; they are specified according to the main canal, the arable land (that is annually inundated land), and marsh-land[40]. Even in the innermost place of the building, the boat-sanctuary, libating Nile-divinities are all around[41], the boat itself being additional evidence of the watery nature of the place. The foundation of this building is undoubtedly the river in its fertilizing aspect, i. e. as the element from which cosmos originates.

It is a *nocturnal* landscape. Night is the other element of chaos primary to a cosmic element: the day emerges from the night. Darkness is a striking feature of the temple interior. The temple is, in fact, like a box. Its landscape lies as if enveloped in night, the darkness intensifying as one moves towards the inner sanctuaries of the building. The pronaos is half-lit through a screen wall[42]; but already the second hypostyle hall is dark, only lit at certain spots through small openings in the roof; the inner chapels lie in total darkness, with no openings in the roof. In this context, the boat sanctuary is a most interesting room. It is, actually, an independent small building, with its own roof, walls, and ambulatory[43]. The sanctuary constitutes the inner core of the temple; the construction of the latter has been compared to that of an onion[44]. When its doors are closed, it represents the centre of the dark, watery chaos.

That the darkness of the interior represents night, is made explicit in different ways, both in decoration and in rituals. The ceiling contains decorative elements identifying it with the night sky[45]; in the ritual texts the god is said to

38 E X pls. CXXI–CXXII; CXXVI–CXXVII.

39 E IX pls. XCIV–CI.

40 Beinlich 1976 pp. 31, 37, 49; as regards the meaning of *pḥw:* p. 28.

41 E I pls. XI–XII.

42 E IX pls. VI, VIII, X.

43 Sauneron and Stierlin 1978 pp. 101, 104.

44 Ricke 1944 p. 11.

45 It is identified with the goddess of the night sky, mother of the newborn sungod, cf. the *wcbt,* E I pl. XXXIII[c], Chassinat room P; cf. too, the vultures in the ceiling of the hypostyle halls, representing U. and L. Egypt – the vulture is a form of the sky-goddess who bears the sungod, Bonnet RÄRG p. 210 f. Also stars can indicate the night sky, as in the chamber of Sokar, E XI pls. CCLXXXI–CCLXXXIII. The ceiling of the vestibule has astronomical scenes. – There is more illustrious material in other Ptolemaic temples, such as the zodiacal figures in the temple of Esna, Sauneron, *Esna* IV 1969 fig. 1; and the stars in the temple of Denderah, Jequier 1924 pl. 61, and in the temple of Philae, Sauneron and Stierlin 1978 p. 157.

"sleep" in the closed sanctuary[46], sleep (and death) being the anthropological parallel to the cosmological chaos.

To understand this chaos-landscape in the temple, it is necessary to consult traditional Egyptian cosmogony. The meaning attributed to chaos according to these traditions is most illuminating regarding the ideological function of the dark temple interior. Chaos is, in Egyptian thought, latent cosmos – hidden in the night and submerged in the inundation waters: it is potential cosmos[47]. The landscape of the temple is identified with this cosmos-in-chaos state; the temple appears as the topos from which cosmos will manifest itself, emerging from its concealment.

The emerging aspect of cosmos is reflected architecturally by a gradual raising of the floor until it reaches its highest level in the sanctuary which contains a boat[48] (plate 8). The sanctuary is the cultic correlate to a mythical place: the place where the solar creator raises himself above the water of Nun to rest on the first mound to appear from the waters, and on which he proceeds to create the world – his light driving the darkness away[49]. This mound – *the high*[50] *seat (st wrt)* of the creator – is represented in the sanctuary[51]; the sanctuary is itself designated The High Seat[52]; on this cultic image of the first cosmic mound rests the boat of the solar creator. With this structural device the innermost core of chaos is shaped into a place which is on the verge of being created. Its shape can be seen as an explication of the cosmogonic implication of chaos.

In the lists of names in the temple there are some which designate it as the first place of cosmos. The building text of the naos calls the temple *The Throne-of-gods of the Gods of the First Occasion*[53]. Another appellation is *Foundation-ground of the Gods of the Beginning*[54]. These appellations refer to the first emergence of the mounds of earth from the subsiding inundation waters; this is explained in the mythological cosmogony texts. When ritual events are correlated with the mythical cosmogony, it is seen that the transition from

46 Cf. the morning hymn inscribed on the façade of this sanctuary, E I 13–20, pl. XXXI[a]; translated and commented on by Blackman and Fairman 1941.

47 De Buck 1922.

48 Chassinat E I pl. II; Sauneron and Stierlin 1978 pp. 36–37.

49 Sauneron and Yoyotte 1959 pp. 35, 46.

50 Kuhlmann 1977 p. 28 f.

51 Sauneron and Stierlin 1978 p. 121; Lange and Hirmer 1955 pl. 243. There is a particularly illustrating picture by M. Audrain in M. Samivel, *The Glory of Egypt*, London 1955 photo 86.

52 Kuhlmann 1977 p. 31 notes 9 and 10. Kuhlmann suggests the translation *hohes Gemach*, referring to its situation on the highest level of the temple floor, p. 32.

53 E IV 1,14.

54 E IV 4,8; VI 6,4.

chaos to cosmos is recurrently enacted in the cult. The cult represents the creation of cosmos as periodical and continuous. When the chaos-imagery of the temple is seen as a constituent of this temple-cultic whole, the intention behind it can be grasped. It participates in a dynamic cultic transition from chaos to cosmos, in which the place of chaos is changed into a place of cosmos[55].

The chaos-landscape of the temple is, then, conditioned by an ontology which sees the relationship between chaos and cosmos as a constructive one. Chaos, the pre-stage of cosmos, is its source of origin, and this source of origin is localized to the temple.

Nevertheless, the transition from chaos to cosmos is not conceived exclusively as a peaceful evolvement out of a state of latency. A parallel cosmogonic tradition of a more war-like character is present in Egyptian thought, where the relationship between chaos and cosmos is understood as being one of antagonism, and cosmos is a victory gained over chaos. Within this tradition the place of beginning is the place where chaos is subdued. The tradition is reflected in the creation-mythology which presents the creator as a warrior fighting and vanquishing the enemy, establishing cosmos through effort – not peacefully arising out of the water to rest on *the high seat,* as in the above-mentioned cosmogony. This tradition is also reflected in the temple ideology of Edfu; the motif of subduing the enemy occupies a central place in the Edfu texts and reliefs. Long mythological texts deal with the fights of Horus[56], numerous reliefs exhibit Horus piercing the enemy with his weapon – the enemy being depicted in various forms[57], and temple names identify the temple with the mythical place of fighting, such as *The House-of-combat (Pr-cḥ3)*[58], *The Place-of-fighting (Bw-tjtj)*[59], *The Place-of-destroying-the-evildoer (St-sẖr-sftẖ)*[60]. Through these means the vanquishing of chaos is localized to the temple.

The motif seems to be polyvalent, *the enemy* being understood in the general sense as that which threatens or is a hindrance to cosmos[61]. Therefore, it has to be defined according to context, but it is apparent that the enemy frequently

55 See part B II.

56 E VI 60–90, 108–136, 213–223. Blackman and Fairman, JEA 21 (1935), 28 (1942), 29 (1943), 30 (1944). Säve-Söderbergh 1953; Junker 1910; Schenkel 1977.

57 E XIII pls. CCCCXCIV–DXIV, DXVIII–DXXXV: E XIV pls. DLXXVI–DLXXXIV. Schäfer 1957; Roeder 1960b pp. 90–154. The motif is found already on the Narmer palette.

58 E VI 176,11.

59 E III 9,9; 33,15; E VI 17,3; 18,2; 329,4; 330,6.

60 E VII 18,9.

61 Hornung 1966 p. 27 ff.

represents the destructive flood threatening the earth and the settlement, the menacing aspect of the inundation which must be pacified before cosmos can be established anew[62]. The majority of the fights of Horus against Seth and his followers evidently has this meaning; the enemies incorporate the dangerous powers of the water in the forms of aquatic animals: hippopotamus and crocodile. They are speared at Edfu, being fought with the harpoon, the attribute of Horus. Timed after the Sothis-star which inaugurates the inundation[63], it is a battle waged in boats[64] on the Nile, from Assuan to the sea, i. e. over the whole of Egypt, following the movement of the inundation.

It is only to be expected that the theme of pacifying the flood opens the cosmogony of Edfu; in the longest record, the subduing of the menacing waters is the first cosmogonic event[65].

Some other definitions of the enemy, relevant within the cosmological imagery of the temple, should be mentioned. The enemy can be specified as the destructive forces of the barren desert, *the red area (dšrt)*, encroaching upon the fertile, *black land (kmt)*, which is Egypt[66]; the desert is the territory which has not been inundated and which cannot be transformed into fertile earth. Seth is the god of the red area[67], and in one of the myths about the combats of Horus[68] the confrontation between these two gods seems to be coloured by these values. The *red* area is also associated with *foreign land*[69], and the enemy has this meaning, too. The latter definition of the enemy is brought into relief on the pylon, which portrays the Horus-king slaying the people of foreign countries – the area beyond the border of Egypt.

The enemy mythologem is thus applied to the inundation when destructive[70], to the earth which has not been inundated, and to the foreign foes when attacking Egypt. Within his framework of cosmological battle, the walls around the temple acquire their meaning as the defense of cosmos against excessive flood, encroaching sands, and invading foreigners. Consequently, the texts

62 Derchain 1978 p. 50.
63 Säve-Söderbergh 1953 p. 27.
64 Cf. the planches referred to in note 57. For the motif in an offering context, see Labrique 1982.
65 E VI 181,11 f.
66 Wb V 126,7.
67 Kees 1943 p. 457, and 1977³ p. 237; Erman 1934 p. 82. The cult-places of Seth were situated at the former borders of Egypt or in the oases in the desert, cf. Hornung 1974 p. 53. According to Plutarch, the priests of Egypt saw in Osiris the power of moisture and in Typhon (= Seth) the power of the dry heat, *De Iside* 364 A, 33.
68 E VI 219–223, XIV pls. DLXXXII–DLXXXIV.
69 Griffiths 1960.
70 Bonneau 1964 pp. 67, 70 f., 102 f.

and reliefs of the myths about the combats of Horus are found on these walls: more precisely on the inner faces of the east and the west walls.

In this connexion a comment on the factual topographical location of the temple should be made. There is a correspondence between the metaphorical meaning of the temple walls as the defense of cosmos, and the pragmatic functions (primarily performed by the outer brick wall). The temples of Egypt were built on a narrow piece of ground between the desert and the river; but as the inundation could – when excessive – ruin the buildings, care had to be taken not to build too close to the river. The walls surrounding them were a protection. On the other hand, since the river was the sole source of water and the main artery of communication, the temples could not be built too far away from it; the temples were, in fact, approached from a quay lying on the Nile or a canal[71]. The site chosen was just above the level of the high Nile[72], on the edge of the desert. The bordering desert was the other enemy, always trying to expand its barren territory, for which reason the temples had to be walled in. The walls of the temples were thus the actual defense against excessive inundation, encroaching desert, and – obviously – invading foes[73].

The mythical enemy can also represent the other element of chaos, i. e. night, incorporating the cosmos-opposing quality of darkness as it captures the light, and over which the god of the sun must be victorious in order to arise in the morning. The temple is the place where He-who-drives-away-the-radiant-one is vanquished[74].

b) Explication of the cosmological imagery

According to the temple ideology outlined above, the dwelling of god is presented as cosmos, topographically and geographically defined as the inundated earth along the Nile. We shall in the following fill in this outline and describe more precisely how this idea is expressed and developed, with the ultimate intention of assessing its theological purport with regard to the concept of god presented by the temple.

71 Badawy 1968 p. 174 ff.

72 Bonneau 1964 p. 101.

73 A particularly illustrating example of the latter meaning can be seen in the reliefs on the walls of the temple of Medinet Habu (20th dynasty).

74 E VI 185,1. The motif belongs to an old Egyptian tradition, see Hornung 1956 p. 32.

1. The source material evaluated

The cosmogony texts

The mythological cosmogony texts of Edfu offer an exposition of the cosmological meaning of the dwelling of god, which is a clue to a more accurate definition of this aspect of the temple. The texts are distributed on the outer wall of the pronaos, on two of its columns, and on the inner face of the enclosure wall[75]. They comprise one long record and several shorter ones, paralleling each other with regard to the main cosmogonic events. The longer record is characterized by an extensive use of glosses, geographical names given to the mythological places created, identifying them with Edfu and its environs; the glosses constitute a detailed definition of the geographical identity of the cosmos created. It is thus the most interesting version from the point of view of this monograph. Also it offers a significant cultic perspective on the creation by employing a ritual mode of creation, namely through invocations and recitals of names. It is for these reasons that it has been chosen as the basis for analysis.

Several questions arise as regards the history of tradition of these texts, some basic ones being the following three. Are they copies of older texts? Are they adjusted and revised versions of older texts? Are they composed in the late period – on the basis of older stock material? These are questions that cannot be answered with confidence at this stage of research. But I would like to mention one factor which goes against the hypothesis that they are mechanically copied from Middle Egyptian texts: the Edfu texts – as well as all Ptolemaic texts – never represent pure Middle Egyptian: they are always of a transitional nature, insofar as they incorporate elements from late Egyptian. If these texts go back to Middle Egyptian originals, they have clearly been revised through the years, a fact which indicates that they have been living, religious realities.

They may differ as to age, though, even if the precise historical relationship between them is not evident. The long text claims to be a copy of a book

75 E VI 181,9–186,10 = E X pl. CXLIX and E XIV pl. DLXI. The text is inscribed on the inner face of the enclosure wall, on the north side.
E IV 358,8–359,8 = E X pl. CV. The text is inscribed on the outer face of the pronaos, on the west side.
E III 7,2–13 (parallels VI 181,11–183,5) and E III 31,11–15 (parallels VI 181,14–15; 182, 2–5). The texts are inscribed on two columns in the pronas.
E VI 14,13–15,11. The text is inscribed on the inner face of the enclosure wall and belongs to the building text of this wall (E VI 5,5–18,15).
E VI 176,3–177,18 = E XIII pls. DXXXIX–DXLV and E X pl. CXLVII. The text is inscribed on the inner face of the enclosure wall.

called *Enumerations of the mounds of the first time* (181,11), and E. A. E. Reymond considers that this book underlies all the cosmogonic texts, regarding them as abstracts from it. She supposes that the book was of general application and not a special work with restricted reference to the Edfu temple[76]. Yet, the long text does indeed have a marked reference to Edfu: its glosses define the cosmic area as a geographical one, implying a cosmogony of particular relevance for Edfu; as these glosses in fact are the very feature which justifies the title of the book, they cannot be regarded as a secondary supplement to it. Nevertheless, the unmistakable similarity between the two texts inside the pronaos and certain passages of the long one, when its glosses are disregarded, seems to indicate that they refer to one particular version of the cosmogony. The long text might be regarded as an elaboration of this version. Alliot, too, seems to have reasoned along these lines, as he suggests that the long text contains a comment on a cosmogony text; it consists of quotations from this cosmogony and glosses explaining their relevance for the Edfu doctrines[77].

The long text may have had a special purpose not shared by the others. Temple-administrative claims may perhaps lie behind; there may have been a wish to establish a detailed Edfu-cosmology that could sanction the domain of the temple. It is interesting to note that the text is inscribed on the enclosure wall, where also the cadastre of the temple is placed (see p. 46): the wall is the place which represents the boundaries of the cosmos of the temple. If this assumption is correct, there is a possibility that the long text may actually have been composed in the late period for this purpose. A special study of the temple-domain of the time, to ascertain if it corresponds with the names listed in the text, might yield information as regards the dating of the text[78]. On the other hand, cultic-ritual purposes may lie behind the glosses. Two factors of significance should be mentioned as regards this possibility. Firstly the text has a marked ritualistic stamp: the creation is effected through invocations and recitals of names. The intention of the text may have been to give a mythological version of the ritual creation through name-recitals taking place on certain cultic occasions (see p. 74 ff; p. 113 ff), with the aim of presenting the mythical authorization of this ritual event by letting *gods* perform the recitals. As regards the identity of such a cultic context it should be noted that secondly, the contents of the text have a close affinity with the cyclical time of creation taking place during the annual inundation; its precise cultic frame of reference

76 MO p. 8 f.
77 Alliot 1966 p. 152.
78 I owe this suggestion to prof. R. H. Pierce.

might be found in the ritually enacted annual creation[79]. Within this context the geographical names would contain the cosmic reality coming into being with – and as a consequence of – the inundation.

Another cultic event has a contextual connexion with our mythological text, namely the coronation of the sacred falcon which actually lived on the temple precincts and was chosen anew every year[80]: On the same wall, close to the cosmogony text, the scenes and texts pertaining to this cultic event are represented. The similarity between the falcon depicted sitting on the mythical resting place in the relief to the cosmogony text, and the falcon sitting on his cultic resting place in the reliefs to the texts, cannot be overlooked. The reason why the representations of the cosmogony and those of the coronation of the falcon are presented on the same spot, is, nevertheless, obscure. There may have been an underlying religious connexion of some kind between the cosmogony and the coronation.

As regards the remaining versions, they share the mythological framework and main motif with the others. The central motif in all of them is the coming of the Horus-falcon to rest on the floater of reed. This motif has, for obvious reasons, a special reference to the temple of Horus at Edfu – a fact which goes against the assumption that the texts might have had a general application and not have been restricted to Edfu. Other elements in the texts have, however, no doubt been widely spread; thus the importance of the inundation in the cosmogonic process belongs to common Egyptian traditions.

Problems of translation

The fact that the texts contain elements of late Egyptian language poses difficulties of grammar, especially concerning the function of particles and the subordination of sentences. The question is whether the scribes utilized the ME tools fully. Adhering to the formalities of grammar this question can be evaded, but if one regards the texts as being religious realities, continually revised to meet the current claims to meaning and purpose, it emerges at once. The question will therefore have to be raised even if it cannot be answered by this author, who hopes an answer will be given by others some time. In the meantime my procedure is to follow the grammar of ME. I want to add, however, that between my interpretation of the contents and my

79 See part B II b 3.

80 E VI 93,2–99,16; 262,11–269,12; 100,2–104,7; 269,14–274,7; 143,12–152,12; 298,2–304,12; 152,14–157,2; 305,2–309,7; 102,3–103,6 (Alliot's order); E X pl. CXLX; XIV pl. 553; Alliot 1944 pp. 561–676; Fairman 1954 p. 189 ff.; 1960 p. 80; cf. Junker 1912 pp. 42–62.

understanding of the grammar there are some unbridged gaps. Probably, some of these will remain unbridgable if one adheres strictly to the procedure stated above.

In spite of these difficulties a fairly coherent picture of the creation can be drawn from the texts. Moreover, when other temple material is correlated, passages which in themselves seem meaningless acquire meaning. It can be demonstrated that the texts belong to a wider temple-ideological whole, and that this context offers clues to the translation.

Another reason why the text is so difficult to translate is its orthography. For many signs there are several possible readings, and often all of them are relevant: It becomes clear as one reads this text that the orthographic polyvalence is intended, as a means of connecting different levels of meaning. The text presents the creation story on different levels of meaning. It is both a cosmogony and a theogony; and this twofold creation is, moreover, presented both as a topographical-geographical event and as a cultic one. The orthography reflects all these aspects of the creation in that an individual sign can be read as different words which can be assigned to the different aspects – the sign, so to speak, uniting them and thus presenting them as pertaining to one, single event.

This refined and complicated method can be seen to be a common feature of Ptolemaic texts, and is not found only in Edfu[81]. I shall not explore the text fully with regard to this capacity for communication, since this can be done better by a philologist; but it is necessary to take note of it on certain points in my translation, as the intended ambiguity of the message is of essential importance for my interpretation of the text.

Have the texts been recited? An extensive use of alliterations indicates that they have; and as the literary category itself can in large parts be characterized as recital (pp. 68–74) and also otherwise shows formal indications of ritual performance, this conclusion lies near at hand. The use of alliteration can of course be a help to ascertain the chosen reading of a sign in a recital[82]. It should be added, though, that a given sign may have different readings, as I have pointed out. There seems, therefore, to be a certain conflict between the ingenuity demonstrated by the pictorial representation of several readings and the restriction to one reading in a recital; and this incompatability might imply that the text had served different purposes – according to whether its message was conveyed as sound or as image.

81 Sauneron, *Esna* VIII (1982).
82 Watterson 1979.

Earlier translations

The texts have been translated by M. Alliot and by E. A. E. Reymond[83]. Reymond has not, however, given a continuous translation but has selected passages. The reader will perceive that my translation diverges in parts from the two previously published. In those cases where the divergencies are due to grammatical interpretation, I have endeavoured to render that in the notes. My grammatical commentary is not, however, a thorough one, like that of M. Alliot. My notes are inserted primarily where my translation needs to be elucidated or where important points in my interpretation are concerned. The text is so difficult that it would exceed the limits of this book if I were to give a detailed philological and grammatical commentary. Nor have I judged it to be within the scope of my task to present a new textual base.

Reymond has also given an interpretation[84] of the cosmogony myth expounding its implications with regard to the temple. The interpretation is inspiring; however, it differs on one particular exegetical principle from the one adopted by the interpretation under consideration, and as this is due to a different evaluation of the literary category in question, a comment has to be made on the interpretation of myth.

The exegetical position adopted by the analysis

The interpretation of mythological texts has never been an easy and straightforward task: it involves particular hermeneutical problems. We shall not here enter upon the general discussion about myth, only consider one insight which has come out of the numerous investigations and debates, and which has been generally accepted. Mythological records do not belong to the category of historical records; they do not direct their attention to the past and bygone, but to the present and actual. When myth states that gods created the world and the temple, it is the contemporary world and the contemporary temple which are given this sanctification, not a world and a temple once upon a time long ago. The cosmogony myth of Edfu deals with the contemporary cosmos and temple – the temple built by Ptolemaic kings.

It is on this point that the interpretation of this monograph deviates from that of Reymond. She interprets the cosmogony myth historistically, on the assumption that it deals with events that took place in very ancient times. On this premise she states that the mythological texts – supported by the building texts – "enable us to gain a fairly clear idea of the development of

83 Alliot 1966; Reymond 1969.
84 MO.

the primitive temples in Egypt"[85]. On the basis of the mythological records and the measurements given by the building texts, she reconstructs the historical development of the temple[86].

Two main objections can be raised to this understanding of the mythological sources. For one thing, it is difficult to see how the reconstruction of the primitive temples can be verified, for the historistic interpretation is not supported by historical evidence, such as relevant archeological material, and thus it remains hypothetical[87].

More important, though, is the hermeneutical objection that can be made against interpreting the texts as if they were historical documents. The time-perspective of myth does not conform to the chronology of history. The time in which the mythical events take place is a dimension of its own, it might be called "mythical time". In Egyptian mythology it is termed *the first occasion (sp tpj)*[88], and *the first time (p3t, p3t tpjt)*[89]; the words denote the time of creation. We would, however, miss an essential meaning of these expressions if we paraphrased them as "past time" or "most ancient time". The time of creation is not the first moment in an eternally proceeding line of moments, the irreversable progress of historical time, but belongs within a time-cyclus according to which there are recurrent *first times* and repeated creations[90]. In our text there are two cosmological, cyclical creations, the one and most important is determined by the annual movement of the Nile, the other is

85 MO p. 258. Her point of view is often expressed in the monograph, and in chapter one she states that "At Edfu the set of cosmogonical records is described as being of divine origin. The introduction to the first Edfu cosmogonical records discloses the tradition that the contents of these records were the *words of the Sages.* We are told that this sacred book was believed to be a *Copy of writings which Thoth made according to the words of the Sages of Mehweret.* The use of the *i3t, mound,* in the plural suggests that this book included, *inter alia,* lists of cultus-places which were believed to have been founded before historical times. It can tentatively be suggested that in this context the expression *i3wt n p3wt tpt,* the *Sacred Mounds of the Early Primaeval Age,* might have been used as a general name of the prehistoric cultus-places of Egypt" p. 9.

86 In an appendix to MO she outlines the history of the Egyptian temple: "The Edfu records, as a whole, are a valuable source-book for the history of the Egyptian temple", p. 323. She reconstructs ground plans of the supposed "Primitive Temple of the Falcon" on the basis of the measurements given in the building texts.

87 Reymond herself draws attention to the fact that no historical evidence is found to support her interpretation.

88 Wb V 278,3; Faulkner CD p. 222.

89 Wb I 496,1.7. – The cosmogony text of E VI 181–185 is taken from a book called *Enumeration of the mounds of the first time, sšrw j3wt njw p3wt tpt* (VI 181,11).

90 Brunner 1955a p. 141ff.; Hornung 1956; Morenz 1960 p. 174ff. Hornung 1966 e.g. p. 13ff.; Asmann 1975b.

determined by the daily movement of the sun. Moreover, it seems as if these two kinds of cyclic creation are brought together in the mythological picture drawn up: the night corresponding with the period of inundated land, and the day corresponding with the period of emerged land.

It might be added that the mythical creations convey a dimension of value rather than of time: the significant characteristic of the creative acts is that they are performed by divine beings. The myth thus focuses on the divine premises of the world of man – again directing the attention to the present and not to the past, although the form of the myth is that of the past[91].

A historistic interpretation of the myth is therefore basically irrelevant and would continue to be so even if it were proven that the myth did contain historical information. Historical facts can be drawn into myth – there are examples of that. But in such cases they are transformed by the value-perspective of myth into something more: in a myth the historical event is sanctioned by sacred authority, the meaning of the historical event lies in its conformity with a model for reality or a sacred prescription. – The prescribing function of the *first-occasion*-events can sometimes be rendered explicit in a way suggestive of ancient times, as when the existing conditions are said to have existed *since the first occasion (ḏr sp tpj)* or said to be *similar to the first occasion (mj sp tpj).* An aspect of chronological duration is unmistakably present in these expressions; but it is, nevertheless, highly unlikely that they refer to time in the modern historical sense[92]: their time-perspective is rather that of "is, was, and ever shall be". At any rate, their significance lies in their value-ascribing function; the accent is placed on the authoritative implication of *since/similar to* the mythical creation[93].

The important exegetical inference from this view is that one should discard a historistic interpretation as being inadequate: it cannot make due allowance for the mythical outlook on life. A most important feature of the creative events set forth by the texts that we are dealing with will be overlooked when they are regarded as taking place in the past, namely their reference to the basic sacred constitution of the contemporary world.

91 See Brunner's distinction between "Urzeit" and "Urgeschichte", 1955b p. 586 ff. See also Luft 1978 p. 6.

92 E. Otto 1966; Hornung 1966; Luft 1978 p. 1 ff.

93 Or, as Brunner has expressed it: "Die Normen des Lebens aber sind, zwar nicht ausschließlich dort, aber vorwiegend, im Mythos niedergelegt. Göttergeschehen ist vorbildlich, verbindlich. Die religiöse Welt aber, der der Mythos zugehört, ist zeitlos. Sie ist jederzeit, in Vergangenheit, Gegenwart und Zukunft, ohne zeitliche Relativierung gültig", 1955b p. 585. – For studies on a comparative basis of the mythical time-perspective, see Eliade 1959 and 1964.

Other sources to the temple-ideology

The mythological temple-texts belong within an ideological whole and should be understood with reference to this. The full scope of the meaning of the cosmogony myth is not rendered explicit in the text itself, but has to be extracted with the help of information given by other kinds of temple texts and temple material. The religious ideology expressed in these texts is also expressed in the ritual events of the temple cult as well as in the architecture and the decoration of the building – the texts are one element in a complex and complete whole. Only when the myth is seen in this wider temple-ideological context can its implications be worked out. The temple itself is the semantically richest frame of reference.

For this reason the cosmogonical texts will be correlated with the *foundation-* and the *building*-texts of the temple, and also with its *donation*-texts. The building-texts are descriptions of important parts of the temple, such as its different chapels, its enclosure wall and pylon. The donation texts are the cadastre of the temple of Edfu.

The reliefs of the temple and its architectural layout will also be drawn into the study; this kind of material reflects the cosmological ideology of Edfu and casts a most elucidating light upon the cosmogony texts.

It should be pointed out, though, that the reliefs are as difficult to interpret as the texts. There are some fundamental questions concerning their interpretation that cannot be answered satisfactorily, because we know too little about the pictorial conventions involved. What is the canonical mode of apprehension in the given instance, that is, the way in which the picture must be seen if its message shall be understood? How are we to determine the mode of apprehending, for instance, the falcon sitting on the top of 3 stalks of reed in the relief to the long cosmogony text (plate 2); do the stalks represent the *ḏb3*-floater which is the *wṯst*-support of the falcon, according to the texts? These designations are written on either side. The question also concerns the available repertoire of the artist: What can he *not* picture? In spite of the difficulties involved in the evaluation of this source material, it can be used, and is of great value.

2. The cosmogony myth of Edfu

The text which this study will concentrate on is inscribed on the inner face of the northern part of the enclosure wall[94]. Partly narrative, partly laudatorial,

94 E VI 181,9–185,2.

and partly enumerating names, it presents a cosmogony which results in the building of the temple of Edfu. The cosmogonic acts have a seasonal dating that links them to the inundation and the sunrise; the temple which is built is the temple of Horus into which the text is hewn, not some primitive sanctuary existing very long ago. The text presents the temple as being erected by divine builders, according to the plans of Thoth and at the command of Horus himself.

The combination of the creation of cosmos and the creation of the temple at Edfu conforms to a pattern which was apparently widespread in the Near East: there are well-known examples from Mesopotamia[95] and Phoenicia[96]. Naturally, this juxtaposition of cosmogony and temple-construction will be given particular attention in the monograph[97].

a. Translation of E VI 181,9–185,2 (pp. XV–XX)

The king: "I have come unto you, Falcon-of-the-mottled-plumage, Horus-of-Behdet, great god, lord of the sky, to bring to you my heart placed on its (right) place, [181.10] (that I may) kiss (you), O living-one[1], at your dromos[2]. For you are the divine god who came into being on the first occasion, at the might of whose majesty the gods kissed the earth!"

A copy of a writing which was made by Thoth, in accordance with [181.11] what the (creative) Words[3] of the Heavy Flood said; *Enumeration of the mounds of the first time,* as it is called.

As to the fighting ⌈...⌉[4], the water [181.12] fell[5].

95 J. B. Pritchard (ed.), *Ancient Near Eastern Texts,* Princeton 1950 p. 60 ff.
96 C. H. Gordon, *Ugaritic Literature,* Roma 1949 pp. 17–27.
97 As regards the relationship between the building of the temple and a concept of cyclical time and creation, see E. Otto 1966, p. 748 f.

[1] Cf. E III 79,4.
[2] *r ḫft-ḥr.k,* Wb III 276,10.
[3] The personified creative words, or those who first utter the creative words: the commanders. See E VIII 108,20.
[4] The text-basis is too weak to carry a reliable translation. But there is an obvious wordplay on the following militant names given to the inundation: *tjtj* associating to *Mw-tjtj* and *Ḥ3/Ḥ3j* associating to *Mw-cḥ3,* which makes it likely that the words refer to a mythical event connected with these names of the inundation. The mythical event in question seems to be that of the primeval fight taking place on the inundated land as the *wḏ3t* (*ad* the inundated land as the *wḏ3t*: see A. Schlott, *Die Ausmasse Ägyptens nach altägyptischen Texten,* Darmstadt 1969, p. 160).
[5] *ḫr wcrt*: The statement implies that the water is subdued – *ḫr* is frequently used in the Edfu-texts to denote the vanquished enemy; it occurs with this meaning also in VI 182,18.

And the name of the flood is
Water[6] -of-fighting,
Water-of-combat.

a)* The two delimitors[7] of a divine heart, united in peace, conducted the

[6] The reading *mw* is uncertain, though likely. The following values of the sign are noted in the Edfu-texts:

a: *jw, island.* The reading is well attested, cf. E III 157,3; 156,1; V 352,13; VI 176,9.

b: *n,* as in *nb ntt,* E IV 358,13, cf. Fairman, ASAE 43 (1943) n. 235; and in *bw pn,* IV 392,13-14; cf. also Dümichen, *Bauurkunde,* ZÄS 8 (1870) p. 6.

c: *j,* as in *jtj,* E IV 32,5, cf. Fairman, ASAE 43 (1943) n. 235.

d: *Jmn, Amon,* E VII 147,15, cf. Fairman, BIFAO 43 (1945) p. 119.

e: *mw, water,* cf. E. de Wit, "Some Values of Ptolemaic Signs", BIFAO 55 (1956), and S. Sauneron, "A propos de deux signes 'ptolémaiques'", BIFAO 56 (1957) p. 77 f.; Fairman, "The Myth of Horus" I, JEA 21 (1935) p. 35; *mw* is the supposed reading of the sign in E VI 128,5-6 (2 ex.), the meaning is obviously *water.*

The choice of value is determined by context. According to the context of VI 181,11-12 two readings might be considered relevant: *jw* and *mw.* The reading of *jw* has much to recommend it: *jw ṯjtj* and *jw cḥ3* would conceptually correspond to *bw ṯjtj* (E III 9,9; 33,15; VI 17,13; 18,2; 329,4.9; 330,6) and *pr cḥ3* (E VI 176,10 f.). The notion of a first island of combat is well-known in the Edfu texts (the *island of raging, jw nšnt* in VI 11,5 might be added).

However, the names enumerated in E VI 181,13 are not given to a place, but expressively to water: *m rn n wcrt.* The reading *jw* will therefore cause a discrepancy between the object, the water, and the names given to it. With the readings *jw ṯjtj, jw cḥ3, jw ḥtp, nt ḏd* only the last-mentioned name is adequate. The reading *mw* is therefore to be preferred.

It might be added that the sign apparently reads *jw* in E VI 182,11; 183,13. There is perhaps an intended ambiguity in the sign as used in this text, to the effect that *mw ṯjtj* and *mw cḥ3* are the preforms, or the chaos-versions, of *jw ṯjtj* and *jw cḥ3*; i. e. the islands are still submerged in water and have not yet emerged as islands.

In Demotic the inundation is termed *mw,* see W. Erichsen, *Demotisches Glossar,* Copenhagen 1954, p. 154 f. See also A. de Buck, "On the meaning of the name *Ḥcpj*", *Orientalia Neerlandica,* Leyden 1948, p. 3 f.

* References to parallel texts are indicated thus. See list p. 42.

[7] *tšwj* corresponds to *nṯrwj, the two gods* in E VI 14,14 and IV 358,13; in both places they are identified with *C3* and *W3j. Nṯrwj* is, though, a word which has a wider application in the Edfu-texts; thus in the foundation texts it designates Re and Horus (e. g. IV 10,8).

It is not clear what functional quality is indicated by the designation, but it lies near to translate *the two delimitors,* connecting the word with √ *t3š, set limits,* possibly with reference to the nome, see Wb V 235.

tš and *wp tš* (E VI 177,6) are technical terms in the cadastre of the Edfu temple, cf. D. Meeks: *wp tš n'est pas relevé au Wb. C'est une expression fréquemment employée dans les textes ptolémaïques, Chou, promoteur du cadastre . . . est désigné parfois par l'épithète wp tš: "Celui qui a délimité les nomes" . . . wp est ici employé dans le sens "délimiter un terrain"* (ref. to JNES 15 p. 30), *Le grand texte des donations au temple d'Edfou,* BdE 59 (1972) p. 55.

Water-of-peace before the *šbtjw*[8], 181.13 *W3j* is the name of one, *C3*[9] the second[10].

In a moment[11] the water stabilized in passing by; the name is Stabilized-water.
And Water-of-fighting,
Water-of-combat,
Water-of-peace,
181.14 Stabilized-water,
became name(s) of the inundation ⌈at the (appropriate) time⌉[12]

A (floating) mass of reed was seen by He-who-is-on-the-water, as the *Ḥtr-ḥr*-falcon[13] was perceived above it, hovering, 181.15 as a wind spread to it.

Alternatively the *tšwj* might be the personifications of the two banks delimiting the area of the high Nile – the area which is later to become cosmos.

[8] In the cosmogonical texts of Edfu the word designates a collective of creators. The meaning of the word is not clear. Alliot suggests *providers* (p. 132 f.); while Reymond suggests *transformers* (MO p. 13); ZÄS 87 (1962) p. 52 f. At any rate, their work apparently has to do with the earth, cf. p. 60 f., and the two *tšwj* therefore are said to arrive *before* them: They are heading the *šbtjw* in rank/in time – as the earth is still covered by the inundation.

[9] *C3* means The Great One and *W3j* perhaps, The Remote One.

[10] Alliot translates *l'Unique* and *le Seul-et-Unique*; the translation is not, however, particularly well suited to a pair of gods – whether conceived of as a duality or as two individual gods. Cf. also E IV 358,13 where the determinative of two persons underlines the concept of two gods. As regards the writing of *The-One-and-Unique*, see e. g. E III 47,9; 42,14; 19,17.

The sign seems to be a graphic variant of the two strokes, cf. E VI 65,7: ⌈*ḥmt*⌉ *2 mn m thn.f*, *the second harpoon is stuck fast in his forehead.* (Professor R. H. Pierce has drawn my attention to the example).

Cf. also C. de Wit, "Apropos des noms...", CdE 74 (1962): the sign of two spears can signify the number of two, e. g. E VII 248,10. Similarly the sign of one spear can signify the number of one, e. g. E VII 18,10; and the sign of three spears can signify the number of three, E IV 14,4; 358,16; VII 15,2.

[11] *cnt cḥct:* Alliot interprets the time-indication as "L'instant de midi (ou: du point culminant = *cḥc*) est celui où le Soleil divin semble suspendu, arrèté en sa course. C'est le moment du règne parfait de *cpy* sur le monde créé en théologie d'Edfou" (p. 133). But as *Cpj*, according to our text, has not come out yet, the time still belongs to chaos, which means, it is night.

[12] Wb I 1,16. It is a debatable question whether *n 3t* is to be construed with what follows or with what precedes; on the basis of content, the latter option has been preferred. The function of the phrase is to indicate that the names of the water are given one at a time according to the different phases of the development of the inundation. For a similar use of *n 3t* see 182,15, where the phrase follows the names given to the town and by concluding this list of names indicates that they are given one at a time in a development of the town. Cf. also E III 7,5. (The suggestion has been offered me by R. H. Pierce).

[13] The meaning of *Ḥtr-ḥr* cannot be ascertained with certainty.

b) On seeing the reed, *W3j* called out (its name). When[14] it was ⌜wanted⌝ in the place, *C3* arrived at the border of[181.16] the district at the edge of the water[15]. When the Beautiful[16]-of-harpoon[17] arrived, the reed separated[18] (from the rest of the floating masses) as it went [19] to them [20]. And a

[14] *jr* appears to introduce a temporal clause (cf. also 181,16; 182,4; 183,1.5), F. Daumas, *Les moyens d'expression du grec et de l'égyptien,* Suppl. ASAE (1952) p. 90. It might be added that this function of *jr* is not unknown either in older stages of the language or in Coptic.

[15] *wcrt, water,* Wb I 288,6; *sheet of water,* P. Vernus, *Athribis,* BdE 74 (1978) p. 235f.

[16] *nfr, beautiful:* A precise translation of the word is difficult to give in our case. Maybe self-renewal or rejuvenation are connoted meanings, cf. G. Jequier, *Considération sur les religions égyptiennes,* Neuchatel 1946; H. Stock, "*nṯr-nfr* – der gute Gott?" (Vorträge Marburg) 1950; cf. too, J. Bergman, in *Actes du 29th Congrès internationale des Orientalistes. Egyptologie,* vol. 1 (1975) pp. 8–14. However, this meaning does not appear wholly sufficient here, as the object is a weapon. An element of dynamic force seems to be connoted by the word; see also E VI 184,6, n. 108. It is to be noted, though, that the motifs of *regeneration* and *fighting* are combined in Egyptian art and literature, and as this combination is especially prominent in our text, both connotations should be maintained.

[17] The second weapon-sign is to be understood as a determinative; cf. E VII 131,12, and also E I 47,15; V 264,13–14; VII 23,4; 35,11. It is the weapon of Horus which is being introduced. Its arrival is connected with the separation of the reed-floater from the rest of the masses of reed; possibly the mythological picture is the floating harpoon – the *ḏb3*-floater being attached to the weapon (Wb V 555,1). The weapon belonging to the seat of Horus is thus joined with this place from its first appearance.

That the weapon is thought to have come with the flood is apparent from various text-references, e. g. E VI 185,15f. where it is called *the august sgmḥ-spear that came out of Nun,* and the parallel-text of E IV 358,2: *Praise to thee, (O) ḥḏ-wr-weapon, the sḫm-ḥr-spear, the august sgmḥ-spear that came out of Nun!*

[18] Lit. *the reed made a delimited side; tš gs: delimited of side.* The situation is that a cluster of reed has separated from the rest of the floating masses of reed; compare E IV 358,14. The topographical connotations of *tš* (see note 7) should be noted – as this segregated reed is identified with the Edfu-area later in the text.

The use of *gs* is prominent in these lines, and there seems to be a wordplay on the sign, connecting the *gs, side* of the reed with the preceding *gs wcrt* and *gs.s* – thus associating the reed with one side of the district.

[19] *ms r.sn:* I construe this as a stative.

[20] Referring to *C3* and *W3j,* cf. E IV 358,13f.; VI 14,14.

floater[21] of reed was stabilized in the water, [182.1] something which the Hovering-one saw while encircling[22]. c) The name of the reed in the water is floater, the protected (object)[23] made by the two[24]. [182.2] The ka arrived at the reed, while this object was (still) being made to move (on the water); the Falcon[25]⌜was (still) in the sky⌝[26]. When the reed was stabilized, the Falcon was supported[27] < by > the floater of reed. The name of [182.3] the floater is Support-of-Horus[28].

[21] The meaning of *ḏb3* is here *floater,* Wb V 555; cf. also A. Gardiner, *Egyptian Grammar,* London 1966[3] p. 514. Wb V 555,1 has noted a special function of the floater which may be of particular relevance in our text: *Der Schwimmer an der Harpune,* as E VI 181,15 appears to give the information that the harpoon, the attribute of Horus, is connected with the resting-place of the god. The combination of 1) resting-place and 2) attribute of the god, i. e. his weapon, is one of the basic motifs of our text, repeatedly re-explicated with regard to Horus-hypostases and protecting activities. The floater in the water, with the harpoon attached, and the falcon hovering above and subsequently resting on it, can therefore be regarded as a model of the sacred place containing the principal three components: the place, the cultic symbol (see p. 98), and the presence of god.

[22] *dbnbn,* Wb V 439; E VI 14,14.

[23] *jḫt pn:* There is a discrepancy between the feminine form and the definite article; see, however, Wb I 124,2.

jḫt pn presumably refers to the attribute of the god, the harpoon, moving down the river with the reed; as regards this interpretation of *jḫt,* see Reymond, "The God's *jḫt*-relics", JEA 53 (1967) pp. 103–106. Reymond offers, though, a very different translation to the passage in question, cf. MO p. 14.

[24] See note 7.

[25] Theoretically, three readings of the falcon-with-flagellum-sign are possible: *nṯr, Ḥr,* and *bjk.* The context decides the choice. As the sign occurs in combination with *wṯst,* in an apparent wordplay on the name of the town, *Wṯst-Ḥr,* the reading *Ḥr* is reasonable. However, Alliot reads *Wṯs-Bjk:* "et il y eut *Wṯs-Bik,* comme nom de cette ville" (p. 139); and Reymond reads *Wṯs-Nṯr* (MO p. 14), but *Wṯst-Ḥr* when the word occurs in the parallel of E IV 358,15 (ZÄS 87 (1962) p. 42.

As regards the other falcon-sign in VI 182,2 the context does not give any explicit clue; the reading of *bjk* is chosen because the text in these lines markedly focuses on the falcon-aspect of the god: the hovering bird is the central figure in the mythological picture – on this apparition of the deity hinges the important event of *the settling* of the god belonging to the sky. Also the parallel texts underline the falcon-aspect of the god in this phase of the cosmogony; E III 31 has the *ḏrtj*-falcon. The falcon-with-flagellum in VI 182,3 might be read *Bjk* or *Ḥr.* As the cosmological implications of the *flying* aspect of the god is explicated here and in the following lines, the reading of *Bjk* has been chosen.

[26] The translation cannot be given with certainty and alternatives could be considered.

[27] *wṯs Bjk* < *jn* > *ḏb3 nbjt.* Alliot, however, reads *wṯs Bik ḏb3* and translates: *le flotteur de roseaux éleva (en l'air) le Faucon,* p. 139.

[28] The reading *Wṯst-Ḥr* has been chosen because of the following identification of the name with that of the town.

And *Ḏb3*-Floater,
and *W̱tst-Ḥr*-Support-of-Horus
became name(s) of this town.
d) “⌈Aurora appearing⌉ alongside the reed[29]!” so said[30] the Falcon.
182,4 ⌈. . .⌉[31].
When[32] the divine Ruler came forth as the *Sj3*-falcon, ⌈the one with a head beautiful of face⌉, the Hovering-one spreading real marvels, making the darkness[33] *m3c*[34] 182,5 with his wings[35], (being) mailclad[36] and (being) ma-

[29] *cnḏj (w) m sw3 gs nbjt,* lit., *Myrrh passing alongside the reed.* Myrrh is here used metaphorically for aurora. The connexion is, moreover, strengthened by a wordplay on *cnḏw, aurora* (S. Sauneron, *Esna* V p. 258), or *sunshine* (Wb I 207,13). As for the reading *myrrh,* see M. Alliot, “Une orthographe non reconnue du mot ᶜNTJW”, RdE 4 (1940) p. 227f.; and for the determinative, see E VIII 74.

The metaphorical use of myrrh is determined by the cultic function of myrrh in Egyptian religion: its perfume heralds the coming god in the cultic theophany. In a similar way the first colouring of the sky before sunrise heralds the coming sun (cf. the following lines). The association between the rising incense and the rising sun is a common one, see H. Bonnet, “Die Bedeutung der Räucherungen im ägyptischen Kult”, ZÄS 67 (1931), where are noted such appellations of the incense as “Duft, der aus dem Horizont kommt” (p. 26). It might be added that the perfume, like the aurora, is not only a sign of the revelation to come but in itself divine presence – being the scent of the god and thus a divine emanation. The wordplay indicates a concurrence of the cultic and the cosmic theophanies. As will be seen, the concurrence of cultic and cosmic events is a dominant feature of this Edfu cosmogony, and more will be said of it later.

[30] The words are spoken by the Falcon. In this interpretation I differ from Alliot (p. 139), and as the allocation of the recurrent *j.n, so said* is of basic concern with regard to the question: who says what?, the grammatical interpretation of the *j.n* is a nuclear one. I understand this clause as *following* the words quoted, as *j.n* cannot be prefixed.

[31] The meaning of *mn ḥntj m jnt ḫndwj* is obscure. Alliot’s explanation (p. 140) to his translation “Que l’espace (infini) reste (à jamais) en retrait!” (p. 139) does not appear adequate.

[32] See note 14. Alliot interprets *jr* as an introduction to a quotation from a text; he considers the glosses to be introduced by *jr: as to, as regards* (p. 152, and p. 130 note a), and translates: Quant à: ’le Seigneur divin apparut, en (forme de) (Faucon)-*si3w’:* (c’est) la tête du... *etc.* (p. 140), the epithets following as a comment to the quotation. This understanding is in accord with his basic understanding of the text as a commentary to passages from a cosmogony text.

[33] *sm3wj*? Wb III 452,7.

[34] Wb II 23. Making the darkness *m3c* means, dispelling the darkness and thus turning chaos into cosmos.

[35] *dm3tj*? The writing is not attested in the *Wörterbuch.*

[36] *pḫr*? Wb I 549,13; II 438,13; cf. J. C. Goyon, “Le cérémonial pour faire sortir Sokaris”, RdE 20 (1968) p. 89. Possibly referring to the multicoloured pattern of the feathers.

jestic, clothed [37] in *s3wj*-gold, (then) "praise!" said the four [38]. "Look!" so said the two gods[39], and so said the *šbtjw*[40]. "Who has come < out of > [41] the underworld?" so said the *šbtjw*.
"The underworld of the ba[42] is the place!" so said the Falcon.
And the Underworld-of-the-ba
became name of this town.
His cry[43] came forth ⌜. . .⌝, 182.7 when the Ruler-of-flying stretched out the ends of the sky, seeing the stabilized Wing[44], ⌜perceiving the first one[45] as his companion[46],⌝ while the Adze[47] was seen fully[48]. As Horus took to flight *(cpj)* 182.8 the Flying Disk *(Cpj)* came into being, the Lord-hovering

[37] *ḏb(3)*, J. C. Goyon, *op. cit.* p. 89; see also H. W. Fairman, *Signs* 266.

[38] Taken to be a reference to the two falcon-apparitions which have already appeared: the Hovering-one or *Ḥtr-ḥr*, and the falcon supported by the reed, together with the two *šbtjw*, *W3j* and *C3*.
As regards the rôle played by the praise and cry of joy when the sun appears, see the Underworld literature and its illustrations, e. g. in E. Hornung, *Ägyptische Unterweltsbücher;* cf. also *ibid.* "Zu den Schluss-szenen der Unterweltsbücher", MDAIK 37 (1981).

[39] Referring to the two falcon-gods who take part in the praise of the outcoming ba.

[40] Referring to the two *šbtjw* who take part in the praisegiving.

[41] *m: out of* has been supplied.

[42] *b3: ba* is a designation of the god in his position of coming out of the underworld; the ba of Horus is identified with the Winged Disk later in the text. For a closer commentary on the concept of ba, see p. 135.

[43] His cry is the cry of joy. As the presence of the star-constellation the Adze indicates that it is still night, the cry has association to the creative cry of the god who emerges from Dat, e. g. E. Hornung, *Das Amduat* I p. 171, see also, E. A. E. Reymond, "Two Versions of the Book of the Dead", ZÄS 98 (1972) p. 126. In this position the creator is described as the god who is heard before he is seen (as he is still in the darkness); and these words are applied to Horus in the parallel text of E VI 15,3 and 17,6.
Alliot translates *pr Nhw gs.f, nty šbn (n) Ḥk3 dbn: (Alors) le (Dieu)-Protégé apparut à côté de lui: c'était (Celui) qui s'était joint (jusque là) au Seigneur-qui-plane (en cercles)* (p. 140f.). E III 31,14 has *pr nh.f.*

[44] i. e. the falcon resting on the *ḏb3*-floater. Cf. the reliefs to the text, plates 1 and 2 (E XIV pls. dlx, dlxi).

[45] i. e. the falcon resting on the *ḏb3*-floater.

[46] *snw*? Wb III 149,8. The translation is tentative, but is well suited to the passage, cf. the following line: "the Lord-hovering coming to give offering to the Lord of *Ḏb3*"; cf. also the relief to the text (E XIV pl. dlxi).

[47] *Msẖtjw:* The *Adze,* Wb II 149. The presence of this constellation indicates that the god has not yet appeared as *Cpj.*

[48] Inaccurate translation of *rsj: at all, altogether.* I expect that what lies behind this difficult passage is that the solar creator stretches out his wings at the point of time when the constellation of The Adze is visible; that is, before he appears as *Cpj.*

coming to give offering[49] to the Lord of *Ḏb3,* elevated as the *Sj3*-falcon of the Place-of-uniting. 182.9 The words passed, and the ba of the Place-of-spreading (-the-wings) heard the name of the Flying Disk. And Horus of *Bḥdt* is the Lord of *Ḏb3,* He-of-*Bḥdt* is 182.10 the Ruler-of-flying, and Ba-of-spreading(-the-wings) is the name of He-of-*Bḥdt*[50].

"Let *pcj*[51]-land appear on it!" so said the *šbtjw.* And the Beautiful-one arrived, Horus-of-praisegiving.

182.11 And Beautiful Island,
and Horus-of-praisegiving
became name(s) of this town.

"Who is our Lord?" [The Wing][52] came, who uttered the word. 182.12 e) And the name passed, being seen? ⌜. . . ⌝[53].

"The name is ⌜He-who-made-the-region⌝[54], the name is Re, the name is Horus-Re[55]", so said the *šbtjw.* 182.13 And the sun united with the heaven upon The Two Lands.

And The-Sun-has-united,
and That-which-is-upon-The-Two-Lands
became name(s) of this town.

"What has arrived at this place, in it[56]?" 182.14 "A high mound! A place of expelling the enemy! New *mspr*-land! Land of He-who-is-great-of-strength![57]
And High-mound,

[49] Literally: *give food into the throat of*; cf. E VI 15,2, where *mouth (r3)* parallels *throat (nfrt): s3ḥ.f rdj.f jẖt m r3 n Bjk nb Ḏb3.*

[50] The town-name of *Bḥdt* is here given the etymology of Place-of-spreading(-the-wings), in a wordplay on *Bw-ḥdd.* 183,15 gives another etymology, namely that of Seat, Throne.

[51] *Ad* the term *pcj,* see Reymond, MO p. 171.

[52] ḏn[ḥ], Wb V 577.

[53] The words are doubtful and I have not attempted a translation. Alliot has suggested: *(et les choses d')en-haut (et les choses d')en-bas seron établies* (p. 143), reading *ḏd ḥrw ẖrw.* This reading seems, however, to be a confusion with *jr ḥrw ẖrw* which is not written like the name under consideration, cf. E III 133,3; V 52,8; 248,14; VII 75,14; 238,8; 267,16; 288,4.
The version of E VI 17,6–7 runs: *ḏ3jsw m snj m ḏdt n k3 n m33tw.f bjk pn smn gs-prw ẖnm jtn nnt tp t3wj: The words passed, uttered by the ka before being seen, this falcon, the ka who founds the temples; and the sun united with the heaven on The Two Lands.* It is tempting to suggest that our unidentified words are parallel to *smn gs-prw.* Also, note that the act is in both cases connected with the cosmic event of *ẖnm jtn nnt tp t3wj: and the sun united with the heaven on The Two Lands.* The name that passes might thus focus on the founding activities of the god. As regards the juxtaposition of cosmic and cultic foundation acts, see p. 52 ff.

[54] *Jr-c*? Wb I 157,14.15.16.

[55] E III 7,2 has *Wṯst-Ḥr-Rc.* In both instances there is a combination of the names of Re and Horus, the latter being identified as the Horus of *Wṯst.*

[56] *jm.s.* J. Osing has drawn my attention to E III 7,3–4 for a similar writing.

[57] Cf. E VI 13,2.

and Place-of-expelling-the-enemy,
182.15 and New-*mspr*-land,
and Land-of-He-who-is-great-of-strength
became name(s) of this town ⸢at the (appropriate) time⸣[58].

f) Horus saw *Mss*[59]. 182.16 "Why?" [60] so said the *šbtjw* who made the invocation when seeing the face of the *r3*-snake:
"Look! ⸢(You) who have united with the pleasing crown⸣[61] drive away your inertness[62]!" 182.17 And when the ka heard the invocation, g) the protector god came out, the image of the Shining-of-face as the *Ḥtr-ḥr*-falcon, the *šm-wḏ3*-weapon in his two hands 182.18 so that the enemy fall ⸢. . . ⸣[63] the Living Falcon[64], the Lord-of-the-head, the God-of-the-temple.

The name of he who opened[65] the water, is the Lord-of-fear. And He-who-opens-the-water 183.1 as the Lord-of-fear[66] came into being. "Look!" the

[58] See note 12.

[59] *Mss* is possibly the deity invoked by the *šbtjw*. The identity of *Mss* is not clear. E III 7,5 has *ḫf Ḥr b3 m Mss: Horus saw the ba as Mss,* which indicates that *Mss* is a ba-manifestation, maybe having a snake-form, as *B3 cnḫ,* the *Living Ba,* is sometimes determined with a snake, e. g. E VI 15,4, see also VI 15,1 and III 7,7–8. The determinative to *Mss* in E III 7,5 is a duckling in an egg.

[60] For *ḥr-m* in this function, see the second Harper's song (Neferhotep II), A. Gardiner in PSBA 35 (1913) pp. 165–170; M. Lichtheim, *Literature* II p. 115f.

[61] *ẖnm ⸢ḫpt⸣ nḏm: (You) who have united with the agreeable crown.* The deity invoked may be identified with the snake depicted as an emblem on the crown of Horus-*sḫm-ḥr* (the Mammisi pl. XIX). There is also a snake on the head of the God-of-the-temple depicted in the relief to the cosmogony text, although the god has no crown. In the legend above him, he is attributed with the *s3-t3*-snake and with the *sgmḥ*-weapon (VI 186,5). The snake thus appears to be an emblem of the protector. It is presumably the counterpart of the enemy-snake; the snake being an ambivalent sign.

[62] *nnj.k*? Wb V 83,2.3. The writing of the suffix is fairly common already in the Ramesside period, see A. Erman, *Neuäg. Grammatik* § 65–67. – The word might have a cosmological reference to the water lying inert on the land, as the immediate consequence of the invocation is that the protector comes as the Lord-of-fear, who opens the canal; the act may have as a function to remove the stagnant and destructive water. The word *nnj* is sometimes used to designate the inundation in this aspect, Wb II 275,7; cf. E IV 103,9.

[63] *ḥr sb3: in the doorway?* The sign of the doorway can, however, also have the value *s,* see H. W. Fairman, ASAE 43 (1943) n. 319.
Has the *šm-wḏ3*-weapon been instrumental in cutting the dyke?
Ad the connexion between the enemy snake and the canal-water, see note 61. The snake is a cosmological and theological preform, see E. Hornung, "Chaotische Bereiche in der geordneten Welt", ZÄS 81 (1956) p. 31f.

[64] *Bjk cnḫ,* cf. E VI 308,2.3.6.7.10.

[65] *ptḥ: open;* the translation is supported by E VI 15,3. Reymond's translation "He-who-created-the-primaeval water" (MO p. 20) violates Egyptian cosmogonical thought. The god can be identified with the primeval waters, but he cannot create them in a *ptḥ*-creation.

[66] Cf. E VII 112,7; III 7,8.

opening[67] (of the water)!" so said the Falcon. When the canal threw it (i. e. the water) out, the Canal *(p3 H̱nw)* is the name of the sacred water[68]; the *r3*-snake [183.2] is the "pupil-of-the-eye"[69] [. . .] It is the Sacred Heart that opened (the water); and the Lord-of-the-opening[70] came; it is he who is elevated[71]. And *Nṯrj* (That-which-is-sacred) and *Ptḥ*[72] (That-which-is-opened) came into being.

[183.3] As for the elevated one coming out, perceived by the *šbtjw,* he is Tanen, the majestic one, coming to the reed, the Support-of-Horus, who lifted up the ancestors, [183.4] and who afforded protection – his strong arm being perceived as the *Wtṯ*[73].

"On the eighth day that which is separated[74] shall be stamped[75], a house of Isden! ⌜. . .⌝" so said[76] [183.5] the God-of-the-temple.

[67] Alternatively: *Look! (the water) has been opened!* The words are elliptical, see E III 7,8.

[68] The term *mw nṯrj* can be applied to any sacred water, whether canal or lake, A. M. Blackman and H. W. Fairman in JEA 30 (1944) p. 16.

[69] *dfḏ,* Wb V 572; cf. E III 7,9. *The eye-pupil* probably refers to the running water of the Nile, see J. de Rouge, "Textes géographiques du temple d'Edfou", RAr 11 (1865) p. 367; A. Schlott, *Die Ausmasse Ägyptens nach altägyptischen Texten,* Wiesbaden 1981, p. 34. If this interpretation is correct, the canalization of the water is seen in terms of a submission of the *r3*-snake, through which event the latter is made an integral element of cosmos.

[70] *Nb ptḥ,* cf. E III 7,10. As regards the meaning of the name: *Lord-of-the-opening,* see note 65.

[71] *ṯn: elevated,* pun on *Ṯnn:* Tanen, god of the earth that elevates from the inundation water.

[72] To be read *Ptḥ* and not *Ptḥ mw,* the water-sign being a determinative; cf. E III 7,10. Both *Nṯrj* and *Ptḥ* are apparently canal-names, possibly applied on the same canal, namely the main canal of Edfu: *p3 H̱nw*; see E. Chassinat, *Le mystère d'Osiris au mois de Khoiak* I, Cairo 1966 p. 209: cf. also E VI 186,6–7 where *p3 H̱nw, Nṯrj* and also *Ptḥ nt* are listed as names of the main canal under its name of *Mw Nṯrj*; see E. Chassinat, *op. cit.* p. 208, and D. Meeks, *op. cit.* p. 33, text 31,9.

The division of sentences is supported by E III 7,10 which has:

sb Nb ptḥ
sw ṯn
ḫpr Nṯrj
ḫpr Nb-ptḥ

[73] *Wtṯ,* Wb I 381; cf. E III 7,11. E VI 325,1 lists *Ḥt-Wtṯ: House of Wtṯ* among the temple-names.

[74] *wḏc, separate, cut out, cut off?* Wb I 404; D. Meeks, AL 2 (1978) 1981 p. 113. The reference is to the area which has been demarcated for a site of building.

[75] *ḫwsj, stamp?* Wb III 248f. The reference is to the preparation of the site for building.

[76] Alternatively *by* (the God-of-the-temple). However, as the parallel of E III 7,12 has left out these words altogether, they belong most likely to the numerous divine utterances which are inserted all through the long text.

h) When the grand Protector reached the *r3*-snake, his father said: "Look! the image of the Protector! Take the *šm-wḏ3*-weapon, 183.6 (you) two arms, and hit! The head (= the head of the spear) is the *Ḥtr-ḥr*-falcon[77] ˹. . . ˺. When 183.7 ˹that which is in the beyond[78] settled down˺, (namely) the *Sj3*-falcon together with the image[79] of the One-great-of-strength, i) the *Sj3*-falcon looked at Horus, powerful-of-face (*sḫm-ḥr)*, (which is) the name of the ˹weapon˺[80]. 183.8 It is elevated[81], the *sgmḥ*-spear[82]˹as˺[83] the Horus-of-the-

[77] The head of the spear has the form of that of the *Htr-hr*-falcon (cf. plate 2). In VI 182, 17 f. *nb ḏnnt, Lord-of-the-head,* and *Ḥtr-ḥr* are both names given to the Protector-god; cf. also E VI 15,4.

[78] *jmjt nfrt, that which is in the beyond,* D. Meeks, AL 1 (1977) 1980 p. 192. Possibly there is a wordplay on *nfrw,* a term for the innermost room of a building, the end or back-part (of a tomb, or human residence): the part of the interior which is farthest removed from the entrance, W. Helck, Urkunden der 18. Dynastie (= Urkunden des ägyptischen Altertums) Berlin 1957 p. 1714; D. Meeks, AL 2 (1978) p. 195; J. Capart, A. H. Gardiner, B. van de Walle, JEA 22 (1936) p. 178. Cf. too, Wb II 260; 261,9. Applied to the temple this part of the interior can be identified with the *Msn*-chapel, which suits our case well, as this chapel contains a falcon-image of Horus and representations of his spear (cf. note 88). As there is a marked tendency in the text to connect the cosmic and the cultic aspect of the creation, both meanings (of *nfrt* and *nfrw*) might be associated – the *Msn*-chapel thus representing the cosmic place where the god from the beyond came to settle. As regards the ontological characteristics of *the beyond,* see the discussion on pp. 104–110.

[79] As has been frequently pointed out, a clear double reference to cosmic and cultic theophanies runs through this cosmogony. Thus the many words for *image of god (snn, ṯjt, sḫm)* associate to the cultic aspect of the creation, see E. Hornung, "Der Mensch als 'Bild Gottes' in Ägypten", in O. Loretz ed., *Die Gottebenbildlichkeit des Menschen,* München 1967. The word *ḫpr* itself, the main cosmogonical term, also has an association to *image (ḫprw)* on the cultic level.

[80] The word apparently denotes some kind of weapon, but I have not been able to identify it. Is it *ṯbt, club, rod*? Wb V 360,13; R. A. Caminos, *Late Egyptian Miscellanies,* London 1954 p. 102. Or, is it *tb3, stick, cane?* P. Posener-Kriéger, *Les archives du temple funéraire de Néferirkarê-kakaï* I p. 67 *Bibl. d'étude* 65 (1976). Cf. also *tb3w,* Wb V 261,7; and *tbj,* Wb V 261,11.

[81] Another meaning ascribed to the name of Tanen. *Ṯn* refers here to the elevated *spear*; and with this application a protective aspect is attributed to Tanen.

[82] Wordplay on *Sj3 < ḥr > gmḥ,* cf. E VI 15,5.

[83] This function of *js* is archaic, see E. Edel, Altäg. Gram. § 828.

sẖm-ḥr-weapon[84] separated[85] (from the water) by Tanen[86]. ⌜"Why is Horus sad?"⌝[87], so said the God-of-the-temple. 183.9 As for the strong arm being seen: As the grand *ḥḏ*-weapon < of > the Maker-of-the-earth was brought to his son, Tanen brought and established the one which is similar[88] before Horus-of-rejoice[89]. (Thus) the strong arm is before 183.10 the God-of-the-temple, the protection of the *šbtjw*.

"Praise and salutation[90] for entering the place-of-building, O, (you) two!" so said the God-of-the-temple. 183.11 And He-who-is-great-of-strength ⌜cried⌝

[84] There is in the passage a vacillation between the god of *Ḥr-sẖm-ḥr* and the weapon of *Ḥr-sẖm-ḥr,* as the attribute of the god can be deified and personalized. This deification of the weapon is more clearly seen in other Edfu-texts, cf. E IV 78,6–7. *ḏd mdw jn Ḥr-sẖm-ḥr, p3 sgmḥ n Bḥdtj: Words to be uttered by Horus-of-sẖm-ḥr, the sgmḥ-spear of The God-of-Behdet*; see also E III 122,2.

A similar fluctuation between god and attribute of god can be observed with *ẖpš-Ḫpšj.*

[85] *sw*: The word is apparently a verb denoting creation, though its precise meaning is not ascertained. It occurs in E IV 358,10: *k3t.sn pw sw jẖt nbt: their work is to sw everything*; E IV 358,18: *r sw jẖt: to sw the things*; E VI 184,11–12: *sw jht t3: to sw the objects of the earth.* Reymond (Jelinkova) has discussed the word in ZÄS 87 (1962) p. 43ff., and offers the following translation of *sw* in these cases: to imbue with or fill with power, or to magnify. In E IV 358,11 the word seems, however, to have another meaning: *sw nn r nf: who separated this from that*; and as this meaning is in harmony with the focus on *separation* as a concept of creation in our text, I have suggested this translation. The image behind the words is the earth (Tanen) rising above the flood again, carrying the weapon – which is thus separated from the water.

[86] *sw jn Ṯnn:* wordplay on *sw ṯn.*

[87] *tp m3st Ḥr ḥr-m,* so Alliot p. 146.

[88] *nḥr,* Wb II 298. The reference is to a spear similar to the one already created. Their names associate with the two cultic spears erected in the *Msn*-chapel of the temple. They are almost identical-looking, both having a falcon-head upon which is fixed the spear-point; one has in addition a solar emblem on the falcon-head (E XI pls. CCXCIV and CCXCVI). The spear with a solar emblem is attributed to Harakhte and has the legend: *p3 sgmḥ n Ḥr Bḥdtj* (E I 232,8); the spear without the solar disk is attributed to the Golden Falcon and has the legend: *p3 sgmḥ šps pr m Nwn* (E I 239,8). These two versions of the *sgmḥ*-spear is, then, differentiated according to the differentiation of the falcon god; see Alliot, *Le culte d'Horus I* pp. 323–325.

Reymond as well is of the opinion that the passage implies that two spears are created (ZÄS 87 (1962) p. 44); but her view is that one is given the name of *sẖm-ḥr* while the other is given the name of *sgmḥ,* which seems to be contradicted by VI 183,7–8 where the two are identified. Cf. also her article "The Origin of the Spear" II, JEA 50 (1964).

In E IV 357,16; 358,2.4 all weapon-names refer to the one mythical spear of Horus.

[89] *Ḥr-3wt-jb:* This Horus-name has apparently conditioned the introductory words to the bringing in of the similar weapon: *tp-m3st Ḥr ḥr-m: Why is Horus sad?* (183,8), focusing on the state of mind of Horus before the second weapon is introduced.

[90] The word refers to the gesture of raising the hands, Wb V 18.

[91]: “Pray, create!” As to that which went out: [92] the order to build, when that which can be tread on is seen, having appeared with the coming of the *šbtjw* 183.12 who proclaim the objects. Momentarily[93], The One-great-of-strength made the invocation, praising the *pcj*-land:

“Mound of the radiant one!”

And Mound-of-the-radiant-one became name 183.13 of this town.

“Island of Re!
That which establishes the earth (= a mound)!”

And Island-of-Re
and That-which-establishes-the-earth
became name(s) of this town.

j) “High mound![94]
B3ḳt-land
That which is powerful of ka!”

And High-mound,
183.14 and *B3ḳt*-land,
and That-which-is-powerful-of-ka
became name(s) of this town.

“⸢Born of water⸣!
Prosperous of seats (of gods)!”

And ⸢Born-of-water⸣ *(Msnt),*
and Prosperous-of-seat
became 183.15 name(s) of this town.

“The seat until eternity,
existing on the top of the mounds which are with the ancestors!”

And the Seat *(Bḥdt),*
and That-which-is-on-the-mounds
became name(s) of 183.16 this town,
the place of the founders since [...]
⸢...⸣ the sages, great of property.

[91] *cš*? (the writing is not noted in the *Wörterbuch*); Fairman, ASAE 43 (1943) sign 318.
[92] The precise syntactic construction of the passage is elusive.
[93] *3t* is used adverbially. For a similar function of *3t*, see L. V. Žabkar, ZÄS 108 (1981) p. 156.
[94] Cf. S. Sauneron, *Esna* V p. 350 c.

183.17 “May you be powerful!
May you be powerful!
O, Horus! as the terror which is behind[95] the high seat, that is there, and which destroys the enemy through its[96] strength!”

And the High Seat,
and 183.18 Destroying-the-enemy
became name(s) of this town.

“The inundation of the earth (*nt s3tw*)[97]!”

And the Throne *(Nst)* [98]
became name of this town,
at the (appropriate) time.

Praisegiving:
“High mound!
184.1 *Mspr*-land of the circumference![99]
High place ⸢-which grows heavily verdant since sunrise⸣!”

And High Mound,
and *Mspr*-land-of-the-circumference,
and 184.2 High Place
became name(s) of this town.

“Enter the place-of-building[100], O (you) *šbtjw*!” so said Tanen. And the *šbtjw* went altogether, arriving at 184.3 the place of settling down. “Look! The place-of-*W3j*!” so said *C3*. “The place-of-the-throne!”
And the Place-of-the-throne
became name of this town.

[95] Referring to the weapon of Horus in the *Msn*-chapel behind the High Seat.

[96] Referring to the *wsr*-capacity embodied by the weapon?

[97] Again we have a combination of the cosmic and the cultic aspects effected through wordplay: *s3tw* has the double meaning of *earth* (R. O. Faulkner CD p. 211) and *floor* (P. J. Frandsen, “A Fragmentary Letter of the Early Middle Kingdom”; JARCE 15 (1978). In our case the floor is that of the temple, representing the inundated earth (p. 11 f.), while the cosmis seat in the water is represented by the cultic seat (p. 13).

[98] The word is given the etymology of *that which belongs to the earth/the templefloor, n(j) s(3)t(w)*.

[99] Cf. E IV 352,15: *šnw n t3*.

[100] Alternatively *Place-of-uniting; Place-of-building* has been preferred on grounds of context. Possibly a wordplay.

⸢"Go swiftly!⸣[101] 184.4 The *twr*-reed comes! Bring the object[102] unto the One-great-of-strength! Look, The *Sj3*-falcon of the place-of-building is in the state of rejoice[103]! May the builders arrive, saluting and praising 184.5 Tanen for the protection (of) Horus-of-the-*sḫm-ḥr*-weapon ⸢before He-who-sees(104] the verdant *twr*-reed⸣!" so said the *Sj3*-falcon, who raises the *šm-wḏ3*-weapon, 184.6 lifted for protection. It is the great *ḥḏ*-weapon[105] that cuts the foliage of bringing harm, the talon[106] of the Falcon, which is spread out[107], its beauty 108] being lowered, 184.7 settling down in the place. The name is the *sgmḥ*-weapon that cuts the foliage[109] of the *twr*-reed. "Indeed, the strong arm 184.8 that ⸢*ḥnk's*⸣⸢110⸣ the *twr*-reed!" so said the Falcon.
And the Strong-of-arm[111] came into being,
and He-who-⸢*ḥnk's*⸣-the-*twr*-reed came into being,
the *twr*-reed which makes verdant ⸢. . . ⸣

[101] *sjn,* Wb IV 38. The word associates with fighting, Wb V 204,10; cf. E I 14,14: *who ran apace and hurled the harpoon at the snout of the Hippopotamus* (translated by Blackman and Fairman in *Misc. Greg.* text E I 17).

[102] *the object* is the attributive property of the *One-great-of-strength,* i. e. his weapon. See Reymond, "The God's *iḫt*-relics", JEA 53 (1967).

[103] The god *rejoices* because the weapon is brought to the One-great-of strength, cf. 183,9.

[104] The reference is to the *Sj3*-falcon, cf. 183,7.

[105] The *ḥḏ*-weapon is here identified with the *šm-wḏ3*-weapon and the *sgmḥ*-spear. In E IV 357,6 it is similarly identified with *sḫm-ḥr,* and in IV 358,2: with *sḫm-ḥr* and *sgmḥ*. In other Edfu texts the *ḥḏ*-weapon can be conceived of as a club, e. g. E IX pl. III; XIV pls. DCLXVI and DCLXVII.

[106] Lit.: *hand.*

[107] Wb III 482,26; cf. E I 306; II 55.

[108] *nfr* has a reference to dynamic force, as it is associated with the falcon as it flashes down with its talon spread out.

[109] *gmḥw, foliage*; the words have apparently been inserted for the sake of a wordplay on *p3 sgmḥ,* it probably does not indicate that only the leaves of the *twr*-reed were cut.

[110] Reading and meaning is uncertain; but as the sentence stands parallel to *The name is the sgmḥ-weapon that cuts the foliage of the reed,* the word may denote a similar act.

[111] *ṯm3-c*? Not attested in Wb. Cf. Fairman, *Signs* 133. The word might be *ṯ3j-c, male-arm,* which would not be quite out of place, as the weapon and the phallos of the god are juxtaposed in the text (184,9f); moreover, the image of the *Sj3*-falcon in the *Msn*-chapel is phallic, and the end of the male member is shaped like a falcon-head (p. 97) – analogous to the weapon with a falcon-head.

"The *twr*-reed has come into being![112]
The verdure has come into being!
⸢. . .⸣ has come into being!" so said the ka[113].

"Horus-of-the-*sḫm-ḥr*-weapon is the god who is in front of the male one, He-of-the-powerful-arm!" so said Horus.
184.10 And He-who-is-in-front-of-the-male-one
came into being,
and He-who-is-powerful-of-arm
came into being.
"In truth, an excellent cutter is with you!"
And Excellent Cutter is the name ⸢. . .⸣.

As to the founders[114]: they 184.11 were seen in front of the builders. Momentarily, the *šbtjw* made the invocation, ⸢the words of the insight of *W3* and *C3*:⸣ "Enter that which has been given as the place in which 184.12 the objects of the earth are *sw*[115], as the water has subsided and the *pcj*-land which can be tread on has emerged!"[116] And the crew of Horus-of-protecting-his-father arrived at 184.13 the place-of-the-water, as the *šbtjw* were saying: "Go to the [quay] [117]! The place-of-the-crew is there!"

The *Ḥtr-ḥr*-falcon had arrived at the place founded as 184.14 the floater of reed. When Horus came, Horus was supported by the reed, and the name of the *pcj*-land is the Support-of-Horus[118], ⸢the *Sj3*-falcon of⸣ the place in 184.15 which he joined his lord[119], the name of which (i. e. the place) is the *šbtjw*[120]. "That which belongs to us is pleasing! Approach!"

[112] Is the implication that not until the *Ḥnk-twr* has come into being and its functional area denoted by a name, does the *twr*-reed appear?

[113] It is unusual for this text to have a *ḫpr* followed by *j.n.* As a similar constellation can be noted in E VI 321,5, I will refrain from explaining it as a copying-error, even if this seems to be near at hand: the passage might have read *twr 3ḫ3ḫ* ⸢. . .⸣ *j.n k3 ḫpr twr, ḫpr 3ḫ3ḫ, ḫpr* ⸢. . .⸣, and would thus be in conformity with the customary pattern of the text.

[114] *ḏdw.* There is a wordplay on *ḏd, stabilize,* connecting the act of stabilizing the *ḏb3*-reed with the act of founding the temple.

[115] Cf. E VI 177,13–14: *njs šbtjw ḥḥ rdjt bw sw jḫt t3 jm: the šbtjw made the invocation: Enter that which has been given, the place in which the objects of the earth are sw,* E IV 358, 10 f: *k3t.sn pw sw jḫt nbt: their work is to sw everything.* As to the meaning of *sw,* see p. 36 n. 85.

[116] The parallel of E VI 177,14 has *sḫt ḥbbt m sj3 bw wḏc jm: when the flood subsided, the place-of-separation was recognized there.*

[117] *dmj* is restored.

[118] The passage apparently gives a juxtaposition of the reed in the water: the place belonging to *Ḥtr-ḥr,* and the reed on the *pcj*-land: the place belonging to Horus, child of *Ḥtr-ḥr* (184,18), see p. 41.

[119] See 182,8–10.

The "reed" was founded[121] there, with regard to a hall[122], the place in which Horus will exist, the place where the ennead is protected, and the seat of His Majesty [... House-of-]duration[123] is the name.

The similar one[124] was made there, ⸢when Horus made his advance, the second one⸣[125] on the arriving of the *r3*-snake, the rod of the One-great-of-strength performing the protection[126].

The crew [. . .]when the *šbtjw* had driven back the water. The crew of Horus arrived when seeing Horus-of-the-*pcj*-land-protected [. . .]. He is Horus, child[127] of the *Ḥtr-ḥr*-falcon. The "reed" was founded[128] when He-who-drives-away-the-radiant-one [129] had fallen [130], as a protection ⸢...⸣[131]. And praisegiving was performed by the crew: "Come! you, who are our lord, 185.2
Harakhte, image of Re! The place is built for you as the place decreed for you by the ka!"

[120] The place is created in their name. For a similar function of the name, see E VI 176,12.

[121] There is a wordplay on *ḏd, stabilize,* connecting the reed in the water with the temple on the *pcj*-land.

[122] The precise syntactic construction of the passage is elusive.

[123] The restoration is supported by E VI 319,8.

[124] *the similar one* is probably the similar weapon which was presented earlier in the text, as the context is again one of fighting against the enemy.

[125] The passage is obscure, but the similar weapon seems to be connected with a second advancing of Horus to meet the *r3*-snake.

[126] Lit.: *protecting altogether (m nḥm rsj).*

[127] *sḏtj,* Wb IV 377.

[128] It is the founding of the temple on the *pcj*-land which is meant. The possibility that the *reed* might be the reed in the water is excluded on grounds of context.

The text draws a distinct connexion, though, between the temple and the reed-floater of *Ḥtr-ḥr,* by designating the *pcj*-land *the Support-of-Horus,* and by stating that the temple "is" the reed (184,15–16), and also, by calling Horus – the god of the temple – descendant of *Ḥtr-ḥr.* As regards the idea behind this identification between the two places, see p. 54 f.

[129] Cf. E VI 17,8. The nature of the enemy seems here to be defined with reference to the darkness, the second element of chaos.

[130] Construed as a pseudo-participle, see H. Junker, *Grammatik der Denderatexte,* Leipzig 1972^{2} p. 109.

[131] *ṯ3w ḥrt.f – the wind of his sky;* or, *ṯ3w ḥrj.f – the wind upon which he is*? I believe the creative dimension of the wind has not been fully explored; nor has it been my intention to do so in this study. For further reference, see S. Morenz, "Ägypten und die altorphische Kosmogonie", in *Festschrift Wilhelm Schubart,* 1950. Cf. also Bonneau 1964, p. 151f.

References to parallel texts

a) 181,12–182: III 31,11–12; IV 358,13–15
b) IV 358,14; VI 177,5–6
c) 181,16–182,3: VI 14,14–15,1
d) III 31,13–15; IV 358,16
e) 182,12–14: III 7,2–4
f) 182,16–183,5; III 7,5–13
g) 182,17: VI 15,3–5
h) VI 15,5
i) VI 15,5
j) 183,13–18: VI 15,8–10

b. Definition of the cosmos created

The text deals with the mythical creation of *cosmos*. It presents the evolvement of a particular landscape, that of the Nile topography, and of a particular geography, namely that of Edfu and its vicinity. The cosmos of the temple of Edfu is thus defined as the very area which pertains to this temple.

1) The topography of the cosmos

The starting point of the creation is the inundation. In this feature the Edfu cosmogony conforms with ancient Egyptian traditions[98]; cosmos evolves out of the watery *chaos* of the Nile flooding the area which is going to become cosmos.

In our text this chaos is described as an entity undergoing a change. It starts by being violent and ends by being peaceful. The change is expressed through the literary device of assigning to the water a series of names which, taken together, represent the dynamic movement. The neutral appellation of the water denotes it as a *sheet of water (wcrt)*. The first qualifying appellation given to this water is the *swirling flood (ḥbbt)*. There then follow names which describe it as hostile and violent: *Water of fighting (Mw ṯtj)*, and *Water of combat (Mw cḥ3)*. These names refer to the destructive aspect of the flooding water, the chaos which is inimical to cosmos. The names given next imply a pacification of the flood: *Water of peace (Mw ḥtp)* and *Stabilized water (Nt ḏd)*. The attributive *peace (ḥtp)* indicates that a decisive metamorphosis of the flood has been effected, it signifies the first step away from chaos. As the word also connotes settling down[99], it can be said to announce the cosmic phase.

98 De Buck 1922; Sauneron and Yoyotte 1959; Derchain in LÄ III 750.
99 Wb III 190 f.

The names form a progression; they describe the water as undergoing a modification. It remains water all the while, but the last-mentioned names make it clear that it has been transformed *from* water in opposition to the state of cosmos *into* water capable of producing cosmos: the place of settling will emerge from it. A war has been waged between the two ontological states[100], and it has ended in the victory of the cosmic forces.

The idea behind the list of names is that the nature of the water is in agreement with the names that characterize it. Here we are in touch with a concept of *name* which occupies a central place in the cosmogony. More will be said of this later.

The aspects of the water denoted by the names correspond to actual phases of the inundation[101].

The subsequent cosmological evolution can be divided into the following main stages: the arrival of the floating reed; the coming into being of the underworld of the place and the stretching out of the sky; the appearance of the first mounds; the opening of the canals; the appearance of the second series of mounds; the clearance of the canals; the work on the land where the mounds are (the *pcj*-land) for the building of the temple.

The arrival of the floating reed is an event of utmost importance. The reed is the first resting place in the fluid element. Again, the topographic image behind the cosmogonic happening is the actual inundation progress: at a certain stage of the inundation great quantities of vegetable matter is suspended in the river which takes on a green hue[102]. According to our text, a part of these masses of reeds separates from the rest and is made into a place of settling for the hovering Falcon: this floater of reed is the first *support (wṯst)* of the god, his perch. It is located at the *edge of the water (gs wcrt),* i. e. the limits of the inundation area, beyond which lies the desert. The edge of the water delineates the cosmological region to come. The *place (bw)* on which the creators *stabilized (ḏd)* the floater *(ḏb3)* is, then, situated on the border between cosmos and the area outside; on this place the temple is erected later in the text.

Next follows the coming-into-being of the *underworld (d3t)* of the place, and its sky is outspanned, i. e. the *ultimate boundaries (ḥntj)* are established[103]. Both events take place while it is still night and only the stars are seen, which

100 The names of *Mw-tjtj* and *Mw-cḥ3* are associated with the initial event through a wordplay of the words *tjtj* and *Ḥ3j*?. The passage is, however, difficult to translate. It starts with *jr tjtj jr*. . . .: *As regards the fighting* . . . or: *As regards the fighting done by* . . . . See note 4 in the commentary to the translation.

101 Bonneau 1964 p. 67 ff.

102 Wiedemann 1920; Bonneau 1964 pp. 65, 291.

103 Compare Brunner 1957 p. 614.

means that from the top to bottom over its whole extent the cosmic room lies hidden in darkness. Then the decisive transition from chaos to cosmos is effected with the coming of the sun. When the sun arises, night is left behind – the one main element of chaos having yielded to the day of cosmos, and the entire space stands revealed. When the Winged Disk appears, all that which the day contains appears simultaneously.

The first mentioned features of the landscape appearing in the light of the day are the mounds *(j3wt)* of the *pcj*-land. They are the solid land which emerges from the water; thereby the other main element of chaos has yielded to the cosmic elements. Again there is a correspondence between the cosmogonical development and the actual topographical conditions during the inundation: when the Nile has flooded the earth, the landscape at a certain stage looks like a collection of islands[104]. On these elevations the houses are built[105]; they are the foundation ground of cosmos, the durable sites of the inhabited world. Later, when the water has subsided, the islands are transformed into mounds connected with the lowland, where the earth is fertilized by the inundation and cultivated to produce the crops which sustain the life of man.

The function of the mounds as the foundation ground of cosmos is given a dominant position in the cosmogony; this is witnessed by the large space devoted to the names of the mounds and by the title of the text, which is: *Enumeration of the mounds of the first time.*

The next important creative operation is the opening of the canal of Edfu. Apart from serving as arteries of communication, canals are indispensible in an agriculture based on inundation[106]. The canals of Egypt are of two kinds, 1: for filling the basins which retain the water when the Nile is low, and 2: for draining the fields, to prevent the water from lying stagnant and spoiling the earth. The opening of the canals is thus a cosmogonic event of consequence. The canalization of the inundation is the final triumph over chaos. In our mythological text it is regarded as a sacralizing act.

That this form of the water is of considerable import in the cosmogony is also evident from the relief accompanying the text: it depicts the figure of Hapi, the personified inundation, as identified with the canal of Edfu; this divinity is placed at the side of the God-of-the-temple[107].

With this act, the area inside the demarcation of the inundation has been turned into a place fit for living. What follows is the completion of the cosmic

104 Bonneau 1964 p. 86.
105 *Op. cit.* p. 102 f.
106 See Butzer 1978 p. iii; Forbes 1955.
107 E XIV pl. DLXI.

landscape, which means the introduction of the last series of mounds and (apparently) the clearing of the canals through the removal of the *twr*-reed. The last-mentioned act implies that some time has elapsed since the opening of the canals. The passage dealing with this episode of cutting the *foliage of the twr-reed (gmḥw n twr)* is somewhat obscure, as it does not define the role of the *twr*-reed. But three features indicate what direction the interpretation should take. One is the fact that it is not the growing of the reed which is the cosmogonic event, but its removal. Clearly the reed is regarded as destructive. This deduction is supported by the information that it is harmful: it is the reed *of doing harm (n wdj nkn)*. The other significant feature is that the protective weapon of Horus is identified with the instrument used in the cutting operation, and this indicates that the reed has a cosmos-threatening function. The third indication lies in the botanical species itself: since reed grows in water, this feature informs us about the place where this harmful vegetation is to be found. It is not a plant that covers the dry earth. If the topographical image of the canal-landscape of the inundated land is maintained – and all evidence points to this reference – the interpretation of the removal of the reed-foliage would be the clearing of the canals[108], one of the most important and characteristic actions maintaining and protecting the agriculture of Egypt. Unless this work is regularly done, cosmos cannot be sustained[109].

The reed in the canals having been removed, the builders of the temple arrive, and the crew of Horus lands at the quay. Then, on the *pcj*-land where the mounds are, the temple is built.

The landscape evolved in this creation myth bears above all the characteristics of an inundation topography. The river winds all through the cosmogony; it is at the beginning as the pre-cosmic element; it changes its destructive qualities and produces the land; it is controlled and made use of in the form of canal-water. The inundation is the source of cosmos, the latter having emerged from it; it upholds cosmos, its lifegiving qualities being perpetuated through the canalization. The cosmos created is the world of canal-agriculture, its origin and maintenance are dependent upon the river.

This dominant position of the water is reflected in the numerous names and designations given to it, presenting ever new modi and functions: *Wcrt* (181,12. 13.15) introduces it as the sheet of water[110]. *Ḥbbt* (181,12; 184,13) refers to it

108 Is there a wordplay on *twr, cleanse* (Wb V 253)?

109 Forbes 1955 p. 54 ff.; Schnebel 1925 pp. 60, 64 ff. It should be mentioned that E II 54,17–55,3, a passage referring to Horus' fighting activities, makes use of a word for *enemy* which literally means *they who ignore his water, wnj-mw.f* (Wb I 314,2.3); cf. also E I 169; 565 76).

110 *Ad wcrt* as designation of the Nile in the inundation time, see Brugsch DG pp. 145 f., 1126 f.

as troubled water. *Mw tjtj* 181,13) and *Mw cḥ3* (181,13) present its destructive aspect. *Mw ḥtp* (181,13) and *Nt ḏd* (181,14) refer to the water in its subdued state. *Nwj* (181,16) and *nwn* (185,8.14; 186,4) relate to its function of being the origin of cosmos. *Ḫnw* (182,1) and *mw nṯrj* (183,1) refer to its being made into canal-water.

It should be emphasized that the chaos-water of this cosmogony is the inundation, not some unspecified, vague primeval water, like that of Genesis I for instance, nor the water of the ocean. The water of the Nile (believed to have its mythical place of origin in the underworld) annually covers that area of the landscape which will, when the flood recedes, emerge as land in cosmos, the place of habitation and the fertile earth that can be tilled. Only where these waters have been can cosmos come into being; it is latently existent in the inundation – its pre-stage and source of origin. The area outside the chaos-cosmos-place is desert: non-world, where no creation can take place. Chaos is the prerequisite of cosmos, and the cosmos of Edfu comes into being on a precisely located narrow strip of inundated land, sharply delimited on both sides by the desert sands.

2) The geography of the cosmos

The text locates the landscape of inundation, reed, islands, mounds, and canals to Edfu. This identification between the mythical landscape and the region of Edfu is effected through the use of glosses: The designations of the mythical places are said to be names of *this town (njwt tn)*[111]. Together, the different names assigned to *this town* trace a precise geographical area.

That the town which is thus the centre of attention in the cosmogony is the actual town of Edfu, is attested by other Edfu texts which refer to the geographical region of Edfu, and which repeat the names from the cosmogony. An example is the donation-text engraved on the outer face of the enclosure wall[112], which contains a survey of the lands donated to the temple of Horus. The text mentions tracts stretching from Gebel Silsileh to Thebes, and it offers valuable information concerning the territory bordering on the canal of Edfu. It has been edited and translated by D. Meeks[113], who takes note of the fact that a prosaic cadastre is engraved on the walls of a temple[114]. From our point

111 Wb II 210.

112 E VII 215–251; XIV pls. DCXLVI–DCLIV.

113 Meeks 1972.

114 "Par leur aspect, leur présentation, les Donations tranchent sur le reste des inscriptions du temple d'Edfou. C'est qu'il est rare, en effet, qu'un document purement administratif s'égare sur les parois d'un édifice sacré", *op. cit.* p. VII. He points to the cultic explanation inherent in the fact that the donation text is presented within an offering context; the products of the earth are offered to the god within the frame of *offering-the field.*

of view this temple-text can be regarded as an element of the cosmological temple ideology according to which the temple represents an actual geographical entity[115]. Nor is it surprising to find included in the donation-text a mythological interpretation of the Edfu reach of the river[116]. It seems logical, too, within our perspective, that the text is placed on the enclosure wall, as this part of the temple represents the boundaries of cosmos, i. e. the place owned and defined by Horus – as is shown by the reliefs on the inner face of the wall, depicting the god slaying his enemy.

Other Edfu texts as well attest the geographical identity of the area of *this town* – making it clear that a real geographical locality is meant. The designations given to the cosmos are, in the cosmogony, identified with names which are documented Edfu-names[117]. Through this identification, mythological etymologies or interpretations are ascribed to the geographical names, and the geographical places are brought into relation with the various regions of the mythical landscape: with 1) the reed, with 2) the underworld of the place, with 3) the mounds, and with 4) the canals. A survey of names documenting this double application to mythical landscape and geographical area follows.

ad 1) *The designations of the reed:*

The designation *floater: ḏb3* is the first name given to the town: *And Ḏb3 … became name of this town (ḫpr Ḏb3 … m rn njwt tn)* (182,3). *Ḏb3* is well attested also outside Edfu as a name of the capital city of the 2nd Upper Egyptian nome[118]. It can also be a name of the nome itself[119]. In our text the meaning of *floater* is attributed to the name. In other Edfu texts other ety-

115 It should be noted that the *offering-the-field* is associated with the sed-festival and the ritual of taking possession of the land by the king, through circumbulation on the temple grounds; cf. Meeks 1972 p. 5 n. 1. See also Frankfort 1969⁶ pp. 85–87.

116 Meeks 1972 text 31, 9, where it says that the *mw nṯrj*-canal came into being in the time of Tanen and lasts until today, *mw nṯrj pw m Wṯst-Ḥr ḫpr.f m rk Ṯnn r mn mjn.*

117 The categories of *geographical* names and *temple* names (see p. 64 ff.) are functional ones. A name is a geographical name when it denotes Edfu; when it denotes the temple, it is a temple name. As regards the historical relation between the two categories some of the geographical names seem to have originated in temple names.

118 Brugsch DG pp. 92 f., 1349 f.; Gauthier DG VI p. 126 f.: "Nom civil de la métropole du IIe nome de Haute-Égypte, – dont le nom religieux etait Behdit". Montet 1961, p. 31 f.

119 E VI 112,5 ff.; transl. by Derchain: Re said to Thoth: "'It means that my enemies are punished *(ḏb3)*'. This nome is thus called Djeba to this day", RdE 26 (1974) p. 13. Cf. also E VII 10,6, transl. by de Wit: "Ce nome est appelé Ḏb3t", CdE 72 (1961) p. 297.

mologies are associated with the name in a similar way, such as *ḏb3: replace, displace*[120] and *ḏb3: retribution*[121].

The designation *support: wṯst* is given as a name of the town together with *Ḏb3: And Wṯst-Ḥr became name of this town (ḫpr Wṯst-Ḥr m rn njwt tn)* (182,3). The name is attested as an Edfu-name by other texts[122] in the forms of *Wṯst* and *Wṯst-Ḥr.* It is, though, the usual name of the nome[123]. According to Gauthier, the form of *Wṯs-Ḥr* generally applies to the nome, while *Wṯst-Ḥr* generally applies to the town[124].

ad 2) *The designation of the underworld of the place:*

Underworld of the b3: d3t n b3 is given to *this town* as the name of *D3t-n-b3* (182,6). The name is an attested Edfu-name[125].

ad 3) *The designation of the mounds:*

a: *The first series of mounds*

The first mentioned appearance is connected with a divine arrival, namely that of *The Beautiful-one: Nfr,* and *Horus-of-praisegiving: Ḥr-j3w.* It entails the assignment of the names *Beautiful Island: Jw nfrt,* and *Ḥr-j3wt* to *this town* (182,11), both attested Edfu-names[126]. The name of *Tpj-t3wj* is given to *this town* upon the uniting of the sun with *the heaven upon The Two Lands (nnt tp t3wj)* (182,13). It is an attested Edfu-name[127].

The designation *high mound: ḥꜥjt wrt* is given to this town as the name of *Ḥꜥjt-wrt* (182,14). It is an attested Edfu-name[128].

The designation *place of expelling the enemies: bw bḥn ḫftjw* is given to *this town* as the name of *Bw-bḥn-ḫftjw* (182,15). It is an attested Edfu-name[129].

120 Wb V 555,6. Cf. Alliot 1966 p. 138 n. f.

121 Wb V 556,6. Brugsch DG p. 922: "le grande ville *Deb*-Edfou, dans laquelle tu as percé les ennemis", ref. to Dümichen *Temp. Inscr.* I 42,2. De Wit, CdE 71 (1961) p. 80: "... c'est la Ville de Rétribution en laquelle la rétribution est exigée de l'ennemi". Blackman and Fairman 1941, p. 409: "Retribution-Town". Derchain, RdE 26 (1974) p. 13: "It means that my enemies are punished. This nome is thus called Djeba to this day".

122 E IV 11,9, cf. de Wit, CdE 71 (1961) pp. 56–97; E VII 2,7; 8,4; 20,2, cf. de Wit, CdE 72 (1961) pp. 277–320. Gauthier DG I p. 210. – *Wṯst:* E VI 198,1.

123 E V 396,1; VI 110,1; 283,13. It is documented as name of the nome since 5th Dynasty, ZÄS 81 (1956) p. 36.

124 Gauthier DG I p. 210.

125 E VI 207,8.

126 *Jw-nfr:* E VI 207,11; Brugsch DG p. 339; Gauthier DG I p. 45. – *Ḥr-j3wt:* E V 396,4.

127 E VI 208,4.

128 E VI 208,7; Brugsch DG p. 560; Gauthier DG IV p. 166.

129 E VI 208,10; E V 396,5; Gauthier DG II p. 33.

The designation *new mspr-land: mspr n m3wt* is given to *this town* as the name of *Mspr-n-m3wt* (182,15). It is an attested Edfu-name[130].

The designation *land of He-who-is-great-of-strength: t3 Wr-ḫpš* is given to *this town* as the name of *T3-Wr-ḫpš* (182,15). It is an attested Edfu-name[131].

b: *The second series of mounds*

The designation *mound of the radiant one: j3t j3ḫt* is given to *this town* as the name of *J3t-j3ḫt* (183,12). It is an attested Edfu-name[132].

The designation *island of Re: Jw Rc* is given to *this town* as the name of *Jw-Rc* (183,13). It is an attested Edfu-name[133].

The designation *that which establishes the earth: ḏd t3* is given to *this town* as the name of *Ḏd-t3* (183,13). It is an attested Edfu-name[134].

The designation *high mound: bw3 ḳ3jt* is given to *this town* as the name of *Bw3-ḳ3jt* (183,13). This expression is not documented as an Edfu-name; however, *ḳ3jt* – which is a technical term in agriculture – is applied to the high-lying terrain of Edfu, withdrawn from the river[135], and it is a component in appellations such as *the highland of Edfu: t3 ḳ3jt Ḏb3*[136], a tract of land situated to the south of the fields.

The designation *b3ḳt-land: b3ḳt* is given to *this town* as the name of *B3ḳt* (183,14). It is elsewhere attested as a name of Egypt[137].

The designation *that which is powerful of ka: wsr k3* is given to *this town* as the name of *Wsr-k3* (183,14). It is an attested Edfu-name[138]. Gauthier has also noted an Edfu-name of *Wsr*[139].

The designation ⌜*born-of-water*⌝: *ms-nt* is given to this town as the name of *Msnt* (183,14). It is an attested Edfu-name[140].

130 E VI 196,11; Gauthier DG III p. 60.
131 E VI 197,4.
132 E V 396,4; Gauthier DG I p. 22.
133 E VI 197,9.
134 E VI 198,9.
135 Meeks 1972 p. 148; p. 56 n. 18.
136 *Op. cit.* text 29,3; p. 100 n. 157. According to Meeks several *ḳ3jt*-lands have probably existed in the immediate environment of Edfu, cf. pl. IV.
137 Gauthier DG II p. 6.
138 E V 396,6.
139 Gauthier DG I p. 20 f.
140 E V 29,1; 396,7; VI 206,4; Brugsch DG p. 299: "nom donné à la ville *Deb*", cf. also p. 1167.

The designation *seat: bḥdt* being on the top of the mounds, is given to *this town* as the name of *Bḥdt* (183,15). The name of *Bḥdt* is among those under which Edfu is best known, also outside the city[141].

The qualification *existing upon the mounds: wnn tp j3wt* is given to *this town* in the form of the name of *Tpj-j3wt* (183,15) which is an attested Edfu-name[142].

The name of *Bḥdt* has here the meaning *seat* or *throne* which is its usual meaning, and this seat is placed within the mythical landscape on the top of the mounds. The following names of the mounds allude to this function of carrying the god's seat or throne:

The designation *high seat: st wrt* is given to *this town* as the name of *St-wrt* (183,17). It is an attested Edfu-name[143].

The designation *throne: nst* is given to *this town* as the name of *Nst* (183,18). It is an attested Edfu-name in the forms of *Nst-Rc*[144]: *Throne-of-Re,* and *Nst-nṯrw*[145]: *Throne-of-the-gods.*

The designation *high mound: ḫcjt wrt* is given – for the second time – to *this town* as the name of *Ḫcjt-wrt* (184,1).

The designation *a high place: bw wr* is given to *this town* as the name of *Bw-wr* (184,1–2). It is an attested Edfu-name[146].

The settlement area (i. e. the site on the mounds) is designated *place of the throne: bw ḥmr* and given to *this town* as the name of *Bw-ḥmr* (184,3). It is an attested Edfu-name[147].

The designation of this settlement area as *the place of W3j: bw W3j* is not explicitly given as a name to the town in the text; it is, nevertheless, attested as an Edfu-name[148].

ad 4) *The canals:*

The meaning *He-who-opens-the-water* is attributed to the canal-name of *Ptḥ-nt* (183,2). It is attested as a name of the main canal of Edfu[149].

141 E V 396,1; Brugsch DG p. 538 f.; Gauthier DG II p. 27 ff., "la ville du trône", nom donné à plusieurs villes d'Égypte qui possédaient des sanctuaires du dieu Horus ... la plus important de ces localités était la métropole de IIe nome de Haute-Égypte ... dont le nom profane était *Deb.*

142 E V 396,6; Gauthier DG VI p. 52.

143 E V 396,1; VI 199,2; VII 1,12.

144 E I 358,16; VI 206,10; Brugsch DG p. 438.

145 E V 396,2.

146 E V 396,5.

147 E V 396,5.

148 E V 396,6.

149 Brugsch DG p. 1166 f.

The usual name of this canal is, however, *The Ḫnw,* which in the cosmogony is presented as the first canal-name: *The Ḫnw is the name of the sacred water: p3 Ḫnw k3 mw*[150] *nṯrj)* (183,1). The *Ḫnw*-canal runs parallel to the Nile. It is the place of embarcation during festivals[151].

Also the name of *Nṯrj,* which is given the meaning *That-which-is-sacred* (183,2) is an attested canal-name[152].

Whether these names refer to different uses of the main canal, or to different sections of it, or to branches of it, is not clear. In the cosmogony the object of attention is the canal-*water.* If the names of *p3 Ḫnw, Nṯrj* and *Ptḥ-nt* designate different canals, they are nevertheless identified with regard to the water they carry; they are all identified with Hapi, the inundation water deified; this is shown by the epithets given to the figure in the relief personifying *p3 Ḫnw* (plate 2): All the names, including that of Hapi, are given to him[153].

As the above survey shows, the mythical places are associated with the geography of Edfu through an aetiological wordplay. The geographical names are understood with reference to a sacred cosmology which centres around Horus; the town with its environment is the resting-place of the god: the *support* which uplifts him; it is the place where he *comes out:* where he has his *underworld*; it is the place where he reigns, having subdued his enemy there: it is the place where his *throne* is established.

It should be emphasised that the cosmos created is not cosmos in the sense of "universe", nor in the sense of virgin "nature". The cosmos which has been created for Horus is the place that is cultivated by man, the irrigation-culture and town-culture of Edfu: a place of Egyptian civilization. The sacred meaning ascribed to the world of Egyptian man in the cosmogony implies the notion that god is present in it. The notion is indicated not only by the geographical identity of the mythological names: it is even more clearly set forth by the fact that the temple – the dwelling of god – is seen as a concretization of this cosmos.

150 *p3 Ḫnw:* Meeks 1972 n. 47. – *mw nṯrj: op. cit.* text 31,9, n. 168; cf. also Brugsch DG p. 379: "le canal sacré" – nom du canal près du temple et de la ville *MASEN* (Edfu) – appelé plus fréquemment *p3 ḫnw.*

151 Chassinat 1966 p. 211; Brugsch DG p. 379; Gauthier DG II pp. 40, 49; IV pp. 173, 196; V pp. 107, 124; Alliot 1944 p. 468 and n. 1; Blackman and Fairman, JEA 30 (1944) p. 16 n. 36.

152 Blackman and Fairman, JEA 30 (1944) p. 16 n. 36.

153 E VI 186,6 ff.

c) The creation of the dwelling of god

1. As presented by the cosmogony myth

The cosmogony culminates with the creation of the temple of Horus; this is what the preceding events have built up towards – the construction of the temple is the final goal of the creation. There are several indications of this.

Firstly, the cosmos created provides the site of the temple; the temple is founded on the *pcj*-land. Secondly, there are explicit references to the planning and the building of the temple dispersed all through the text. The most significant ones are as follows:

The intention of building the temple is first uttered when the canal of Edfu has been opened. Immediately following upon this event the God-of-the-temple declares: *On the eighth day that which is separated shall be stamped, a house of Isden (r3–5 + r3–15, ḫws wḏc ḥt Jsdn)* (183,4). Two important points connected with this order should be noted, one being that it is given directly after the opening of the canal, the other that the temple is designated *a house of Isden,* that is: a house of Thoth, Isden being a Thoth-name. As regards the first-mentioned point, the declaration is well-timed. Only when there is a road of communication can the temple be built; the building operation requires access to the river, without which the material cannot be transported. But the site of the temple is, as we have seen, on the edge of the high inundation, which means that it lies away from the river when the river is low. The canal offers the required access to the river. It might be added that certain building operations required high water in the Nile and the canal leading to the site, i. e. when heavy stones were transported[154]. As regards the second point, the appellation *house of Isden* implies that the temple is designed by Thoth, and the appellation implies that the temple is in the planning-stage. The role here assigned to Thoth is a traditional one[155]. The plans of the temple are under the auspices of this god; he sets up the architectural principles and the decorative rules. Ptah is the craftsman god[156], and is the traditional executer of the plans of Thoth. The wordplay on *ptḥ* made by our text in connexion with the opening of the canal – it is *opened (ptḥ)* by the *Lord-of-opening (Nb-ptḥ)* – may allude to this role played by Ptah as the divine temple builder: Ptah having arrived on the scene, the house planned by Isden can be built.

154 Müller 1967 p. 358 ff. Further, H. Chevrier, "Techniques de l'Égypte antique", *Acts du XXIXe congrès intern des Orientalistes,* vol. I p. 30 ff.

155 Boylan 1922 p. 89 f.

156 Mariette, Dendérah III (1871) pl. 70; Sandman Holmberg 1946 p. 45 ff.

The words uttered by The God-of-the-temple thus place the opening of the canal in a temple-building perspective. The words of the god are no digression, but state the *raison d'être* of the canal from the point of view of the temple construction.

Before and after this episode there are passages dealing with the coming into being of the weapon of the Protector. At first sight they might seem out of place. However, in our text the weapon is conceived of as that which protects the sacred place, and to all appearances it is introduced in this capacity already from the beginning; it comes with the reed, attached to this first resting-place of the Falcon as if fastened to a floater, the place and the weapon belonging together: the reed-floater is *protected* (182,1). The coming of the floater and the harpoon introduces the theme of subduing the enemy from a cultic point of view. The sacred place is by definition a place protected. The actual relevance of this characteristic as regards the temple is evident. The temple is the place where god has vanquished his enemy; this is reflected in the names of the temple, and in its narrative records and reliefs. It may be added that the weapon also has an important cultic position, as it is set up in the first chapel of the temple, the *Msn*-chapel, which lies behind the *St-wrt,* and which contains images of Horus as the *Sj3*-falcon and as The Golden Falcon. The weapon of Horus is here represented by an actual harpoon, possibly two[157]. The temple is, like the mythical resting-place of the reed, a place protected by the harpoon.

When the order to build goes out to the builders (183,11), it has as its immediate consequence the finishing of the *pcj*-land, which means that the last series of mounds is created. The temple is going to be erected on the *pcj*-land. Moreover, several of the names conferred upon these mounds allude to their functioning as a seat or throne of the god:

Wḏ3-st:	*Prosperous-of-seat* (183,14)
Bḥdt:	*The Seat* (183,15)
Tpj-j3wt:	*That-which-is-on-the-mounds* (183,15)
St-wrt:	*The High Seat,* having as an additional name *Destroying-the-enemy,* because the enemy is destroyed there (183, 17–18).
Nst:	*The Throne* (183,18).

These are names associated with the temple built on the mounds. The temple is presented by the text as *the seat of His Majesty* (184,16) and as a *place of protecting the Ennead* (184,16), designations echoing the combination of seat and destruction of the enemy.

157 See p. 97 f.

The *šbtjw* are the creators of the site on the *pcj*-land, but the actual founding and building work is effected by *founders (ḏdw)* (184,10–11) and *builders (ẖnmw)* (184,4). What the *šbtjw* actually do, is obscure. They are reported to *sw the objects of the earth there (sw jẖt t3 jm)* (184,11–12). A more detailed discussion of their activities will be given in a later chapter[158]; here it will suffice to say that their work apparently involves a kind of creation, though the meaning of *sw* is not known. It seems as if the task of the *šbtjw* is to cultivate or prepare the earth in some way or other.

An implicit reference to the connexion between the cosmogony and the temple construction lies in the fact that the temple appears as a kind of duplication of the floater of reed. There are many links between the reed and the temple to witness this:

The site prepared by the *šbtjw* is identified with the *place of W3j (bw W3j)* (184,3). This means that it is situated at the place where *W3j* stabilized the floater of reed, i. e. at the edge of the inundation (182,15–16); it is now, however, lying on the mounds of the *pcj*-land, as the water has subsided *(cẖmm jn ḥbbt)* (184,12), leaving the place of *W3j* at the border of the desert at this stage of the movement of the river.

In other ways, too, the temple built on the *pcj*-land is associated with the floater in the water. For one thing, the name of *Wṯst-Ḥr (The Support-of-Horus)* is given to the *pcj*-land (184,14). Further, the act of founding the temple corresponds to the act of stabilizing the floater. This appears from the words stating that the "reed" was founded *(ḏd)* on the *pcj*-and – with a hall *where Horus will exist,* a place *where the ennead is protected,* and a *seat of his majesty.* In other words, the *reed* is here the temple. Finally, Horus, for whom the temple is built, is called *child* or *descendant* of the *Ḥtr-ḥr*-falcon, thus being related to the Falcon that settled on the reed.

The precise relationship between these two supports of the god, the reed in the water and the *reed* on the land, can be explicated in the following way:

1: The site on which the temple is built is located where *W3j* stabilized the floater of reed. Owing to the progression of the inundation, the place is not, however, in the water any longer, but on solid ground; the temple is founded on the mounds of the *pcj*-land after the flood has receded. There is, nevertheless, an emphasis on the fact that its site is the *site* of the reed in the water, it is *the place of W3j.*

The elevation of the earth is, according to our text, the result of the coming of the earth-god to the reed: *As for the elevated one coming out, recognized by the šbtjw, he is Tanen, the majestic one, coming to the reed, The Support-of-*

158 See p. 60 f.

Horus (jr ṯn bs m sj3 jn šbtjw sk Ṯnn pw šfjt s3ḥ nbjt Wṯst-Ḥr) (183,3). Paraphrased, the earth emerged at the place of the reed.

The temple, then, *succeeds* the reed. It is placed on the same spot where the reed uplifted the Falcon.

2: The temple *represents* the reed: The temple exercises the function of supporting the god. Its site, the *pcj*-land, is said to be The Support-of-Horus, and it contains in itself the seat of the god, the place where he is uplifted above the water – the latter being represented by the temple floor[159].

Also in another respect the temple has a representational similarity with the reed: the temple is protected by the weapon of the god, the weapon which was attached to the floater of reed.

The relationship between the reed and the temple is, then, that between a model and its replica, the model formulating the function and the nature of the sacred place: It is the support of the god and a place protected by his weapon.

3: A third interpretation of the relationship between the support in the water and the support on the land, might be possible. The two supporting places may refer to two distinct sanctuaries, located on different places: one being a quay-sanctuary, the other being the main temple – both carrying the name of *Wṯst-Ḥr*[160].

Nevertheless, the evidence given by the text under consideration does not agree with this interpretation. Firstly, there is the fact that the resting-place of *Ḥtr-ḥr* is identified with *this town,* an entity which, as the cosmogony evolves, turns out to comprise an extensive area, including mounds and high-lying regions – and most expressly so – not only the region near the water. Secondly, there is the fact that the main temple is built on the place of *W3j,* when *the water has subsided*; thus the two sites coincide.

159 See pp. 13, 38 notes 97, 98.

160 It is an attested fact that the names of *Ḏb3* and *Wṯst-Ḥr* are applied to the main temple of Horus. The existence of a second and minor temple called by these names is a conjecture, based on Alliot's interpretation of texts pertaining to the festival of the union between Horus and Hathor (1944 pp. 454 f., 458): The boats coming from Denderah first reached Edfu at a sanctuary lying at the landing stage. At this sanctuary the statues of the gods were taken out of the boats and uplifted *(wṯs)* onto the litters that took them to the main temple. The place was called *Wṯst-Ḥr, The Litter-of-Horus,* at the arrival, and *Ḏb3* at departure, when the images were taken out of the litters to reembark. According to Alliot *Wṯst-Ḥr* and *Ḏb3* designed the area lying north of the urban centre of the town of Edfu, while *Bḥdt* designated the main temple of Horus and the quarters around it, to the south of the ancient city *(op. cit.* p. 458).

It is difficult to avoid the conclusion that the temple represents the floater of reed. It performs the function of the latter of supporting the god; it is a place protected by the weapon of the god. Also, it has a similarity to the reed: as has been shown in an earlier chapter, its floor is the inundated soil from which "reed grows" in the form of a cluster of pillars: a *place-of-the-flood (bw ḥbbt,* 184,13); and above it the falcon hovers engraved in the ceiling or on the architraves: uplifted above the temple landscape.

This relationship of representational identification is attested also by the facts that the ritual foundation of the temple connotes the cosmogony, and that *Ḏb3* and *Wṯst,* and other names of the mounds on the *pcj*-land are documented names of the temple of Edfu[161].

2. The ritual foundation of the temple connotes the cosmogony

The foundation of the temple is represented in texts and reliefs depicting acts performed pertaining to the laying out of the site and its preparation. Together with the building texts, this material presents the foundation in a manner which points to its being conceived mythologically. The operations are seen in the light of the cosmogony, by way of letting the king – founder and builder – carry out the work assisted by divinities known as creators of cosmos, and by including allusions to cosmogonical events. The king represents in this role the will of Horus and is called son of the god. That the Ptolemaic king is shown working in co-operation with the creators of the world is one of the numerous indications of the actuality of the myth. The mythological aspect thus applied on the foundation and building operations brings to light their religious meaning; it has the effect of ascribing a cosmogonical meaning to the founding and building the temple of Edfu.

The reliefs depicting the foundation of the temple are stylized and simplified, concentrating on a choice of acts performed by the king. It is a generally accepted opinion that they refer to a foundation ritual[162]. The repertoire comprises 10 scenes, which are not, however, always presented in total[163].

The most frequently represented act is *stretching the cord (pḏ šs)*[164]; the scene functions as a kind of vignette to the foundation acts. In the reliefs the

161 See ch. 3 on the nomenclature of the temple.

162 Montet 1964.

163 Montet 1964 p. 75.

164 E II 31 = XII pl. CCCLXIX; III 105; 114; 167; IV 352 = X pl. CV; VI 168 = X pl. CXLVII.

king is shown measuring the site together with Seshat, goddess of writing and calculation, a kind of female duplicate of Thoth. At Edfu she is called *Mistress of the ground plans and the writings*[165]. Sometimes Thoth is also depicted taking part in the act, standing behind Seshat and noting down the result of the measuring[166].

The other principal acts shall be mentioned briefly[167].

Hoeing the earth[168]: With a mattock the king digs the foundation pit down to the ground water which is called Nun[169] – a name of the mythical primeval water, having strong cosmogonical connotations. Like the *d̲b3*-floater, the temple of *D̲b3* is founded *(d̲d)* in the primeval water.

Moulding the bricks for the four corners of the temple[170].

Emptying a bucket of sand[171], an act representative of the lining the foundation pit with sand.

Presenting 17 plaques to Horus[172]: The plaques are small and made of silver, gold, copper, lapis lazuli, turquoise ... and are representative of all precious metals and stones needed for the temple. They are apportioned into four deposits on the four corners of the foundation ground. Other things are also put into foundation deposits: amulets, tools, and models of tools[173].

Moving a block of stone[174]: The king is shown with a lever pushing a block of stone; the act is representative of the building operation and the title of the scene in one of the instances is: *Building the High Seat of Re from the beginning of stone shining and hard*[175], thus applying the mythological category of *the beginning* to the actual temple.

Purifying the building ...[176]: The king is depicted purifying a small naos with grains of natron.

165 E II 31,6.
166 E IV 352,12 ff. = X pl. CV; VI 168,15; 174,7 ff. = X pl. CXLVII.
167 The acts are arranged in the sequence propounded by Montet 1964.
168 E II 60 = XII pl. CCCLXXI; III 106; 166–167.
169 E VIII 237; see Brunner 1957 p. 619.
170 E II 60–61 = XII pl. CCCLXXII; III 114; VII 48. The purpose is obscure; see Montet's suggestion, 1964 p. 89. The rite may belong to an old tradition and refer to an act which has lost its practical significance.
171 E II 3–2; 107; VII 46.
172 E II 32 = XII pl. CCCLXXV; VII 47.
173 Montet, *op. cit.* p. 93; RÄRG p. 263.
174 E II 61 = XII pl. CCCLXXVI; III 107 = IX pl. LXI; VII 49.
175 E III 107,14 = IX pl. LXI.
176 E II 32–33 = pl. XL b; II 62 = pls. XLC, CCC xlxxvii; III 109; VII 50–51.

Offering the building to the god ...[177]: The king presents a model of a naos in his hand to Horus. The picture may refer to an actual offering act, the general assumption being that all scenes depict ritual acts.

These texts, then, give a summary presentation of the construction of the temple, and thus they are a supplement to the cosmogony text which refers to this work without specifying it.

The regularly occurring divine co-workers of the king are Seshat, Thoth and his hypostases Isden and Tekh, Tanen, the *ḫnmw*-builders, the *ḏ3jsw*-commanders and the *šbtjw*.

Thus, in the building-text of the naos, Seshat measures the site together with the king; Tekh directs the work, the *ḏ3jsw*-commanders *sanctify the objects* of the temple *(ḏsr jḫt.s)* and the *ka makes its formulas great (swr mdw.s)*, i. e. he recites its magnifying praise; and Tanen founds the temple[178].

The foundation text on the inner face of the enclosure wall mentions Seshat who measures the site, Thoth who plans the temple, Isden who inscribes it, Tekh who gives the instructions, the *ḫnmw*-builders who found the four sides, the *šbtjw*[179] who give the praise *(j3w)*, and – later in the text – the *ḏ3jsw*-commanders who perform its magnification *(swr mdw)*[180].

The foundation text on the outer face of the enclosure wall mentions Tekh, Seshat, the *ḫnmw*-builders and the *ḏ3jsw*-commanders *who magnify the foundation wall (swr snṯ)*[181].

The foundation text on the exterior of the naos mentions Seshat who performs the measuring act, Thoth who gives the dimensions, and the *ḫnmw*-builders[182].

The building text on the outer face of the enclosure wall mentions Seshat, Tekh, the *ḏ3jsw*-commanders, and the *ḫnmw*-builders[183].

These gods working together with the king, who represents Horus, lord of the temple, are all – with the exception of Seshat – gods to whom cosmogonic functions are ascribed by the cosmogony text. Their partaking in the founding and the building of the temple of *Ḏb3* is one of its conspicuous links with the cosmogony of *Ḏb3*. I shall give a brief account of their twofold undertaking.

177 E II 33; 62–63 = pl. CCCCXXVIII; III 110, pl. LXI; VII 56–57.
178 E IV 14,7–9.
179 E VI 170,2.
180 E VI 175,2.
181 E VII 49,4–9.
182 E IV 352.
183 E VII 6,1–2.

Even though *Seshat* has no particular role in the cosmogony, her measuring act has an unmistakably cosmic mark, as its performance is timed and directed by a special star constellation given an analogous function in the cosmogony, namely the *Msḫtjw*[184]. Before the foundation the stars are examined and the measuring is done with reference to *Msḫtjw*[185]. In the long cosmogony text the *stretching out (pḏ)* of the *utmost ends (ḥntj)* takes place when this star constellation[186] is seen. There is at this point a correspondence between the laying-out of the cosmic area *(pḏ ḥntj)*, and the laying out of the temple site *(pḏ šs)*. As the utmost ends of cosmos are stretched out (with the wings of the Ruler-of-flying) and the limits established while it is still night, so the cord is stretched over the foundation site and its sides are established while it is still night.

Thoth is central in the cosmogony. He has made the book containing the specification of the mounds on the *pcj*-land[187]. His role in the foundation and building of the temple is that of the chief planner; he is god of the ground plans and the construction prescriptions[188].

Tanen is god of the earth which rises from the water, a characteristic which in the cosmogony text is reflected in the wordplay on *ṯn-Ṯnn*[189]. On this elevated earth the temple is founded. Expressed in another way: *Wṯst-Ḥr* is said to have come into being *in the time of Tanen*[190]. The concept of time is here a qualitative one, as it is the content of the period which is underlined by the appellation; the period is named after the god who reveals himself during it. The time of Tanen is the time when the creative properties of the god are displayed, when the god has emerged and shows the quality of his being.

In addition to his functions as *earthgod,* there is the possibility that Tanen is identified with *Ptah* in these contexts. Ptah is the traditional craftsman god, the god who realizes the plans of Thoth[191]. He is also the traditional bringer of the inundation[192], and this cosmogonical function seems to be reflected in the Edfu cosmogony: the god who opens the canal is surely a Ptah-manifestation; the play on the word used to denote *open: ptḥ,* has associations with

184 E VI 182,7.
185 E III 167,15; VI 14,8; VII 44.
186 E VI 182,7.
187 E VI 181,10–11; cf. Boylan 1922 p. 89.
188 Boylan, *op. cit.* p. 89 ff.; Bleeker 1973 p. 142.
189 E VI 183,3.8.
190 E VI 326,1.
191 Boylan 1922 p. 91. The mythical architect of the temple, Imhotep, is called *son of Ptah* (E VI 10,10).
192 Bonneau 1964 p. 232 n.1.

the name of *Ptḥ*. This suggestion is confirmed by an appellation expressly given to Ptah in another Edfu inscription: *Ptḥ nt*[193]. Nevertheless, the cosmogony text identifies the opener of the canal with Tanen, who is given the prominent place as the god of the land emerging from the water. It seems as if Tanen here executes his task as Ptah, and a corresponding identification between the two gods may lie implicit in the foundation texts[194].

The creator collectives

The *ḏ3jsw*-commanders operate both in the cosmogony and in the foundation of the temple. They are the personified creative words, and as such they are presented in the long cosmogony text[195]. They are from the time of *The Heavy Flood,* the precosmic phase when water is the dominant element; and their *going out* calls cosmos into being[196]; through the commanders the transition from chaos to cosmos is effected. Their creative function thus occupies a key position in the cosmogony of Edfu[197].

The *ẖnmw*-builders are presented as the builders of the temple in the foundation texts as well as in the cosmogony text. They are not, however, without cosmogonical affinity, since they are called *offspring of Tanen*[198], a designation which alludes to their status in cosmos, and they are reported to constitute – together with the *ḏ3jsw*-commanders – the company of *The Thirty Gods*[199], the comprehensive collective of the creators of the world.

The cosmogonic functions of the *šbtjw* is connected with the *pcj*-land. The *pcj*-land comes into being on their order: *Let the pcj-land appear on it! so said the šbtjw* (182,10). In the cosmogony attached to the foundation text on the exterior of the naos they are presented as those who *bring into being (sẖpr)* the *pcj*-land of every Great Place[200].

The word especially used to denote their creative work is, however, *sw*. Its precise meaning is not known. In a presentation of the *šbtjw* it says that the purpose of their work is to *sw all things (jẖt nbt): These great and eminent gods, the noble šbtjw in Ncrt, the providers are they, who provide, the eminent*

193 E VII 112,7.

194 The identification between the two gods has long traditions in Egyptian religion, see Sandman Holmberg 1946 ch. IV.

195 E VI 181,10–11.

196 *Ad* the expression *ḏ3jsw snj,* see E VI 182,9, cf. VI 182,11–12; VI 17,6.

197 See p. 72.

198 E III 317,13.

199 E VI 173,6 f.

200 E IV 357,17.

201 E IV 358,10 ff.

of the eminent ones, their work is to sw all things (nṯrw jpn c3w wrw šbtjw špsj ḫntj Ncrt ḏb3w pw ḏb3 wrw n wrw k3t.sn pw sw jḫt nbt)[201].

All things (jḫt nbt) seems to be synonymous with the *things of the earth (jḫt t3)* in the long cosmogony text: *the šbtjw made the invocation, the words of the knowledge of W3 and C3: Enter that which has been given as the place in which the things of the earth are sw (njs šbtjw ḏ3sjw m sj3 W3 C3: ḥḥ rdjt m bw sw jḫt t3 jm)*[202]. The *šbtjw* make a similar call in the cosmogony connected with the foundation text on the enclosure wall: *Enter that which has been given, the place in which the things of the earth are sw (ḥḥ rdjt bw sw jḫt t3 jm)*[203]. We might tentatively give the translation *created* in these instances. Possibly the word denotes a separation[204].

What the identity of the *things (jḫt)* or *the things of the earth (jḫt t3)* is, constitutes a problem in itself; but the presentation of the *šbtjw* quoted above might give a clue: the work done by them appears to affect the *products of the earth,* as they are called *the providers who provide*[205]. The *sw* may refer to some kind of cultivating work on the land.

This creative work on the *pcj*-land, whatever it has consisted of, is the cosmogonic function assigned to the *šbtjw*. It may – as is the case with the work of the other creators – have a correlative in a foundation act and be mythologically related to the preparation of the temple site. Reymond connects the cosmogonic *sw jḫt t3* with an act of *sanctifying the objects* which apparently took place at the laying out of the site: in the building text of the naos there is a *Stretching-the-cord*-episode which states that Seshat loosed the cord, Tekh directed the rites, the Ogdoad[206] marched around the site, and the *ḏ3jsw*-commanders *sanctified its objects (ḏsr jḫt.s)*[207]. As the *ḏ3jsw* represent the creative words, the sanctifying act must have consisted in the recital of sacred formulas. Reymond next maintains that *sw jḫt* stands parallel to *dm jḫt* and that *sw* has an analogous meaning to *dm: to name*[208]. On the basis of these observations she interprets *jḫt t3* as sacred objects "which symbolised that which was to be brought out of the water", and she sees the *sw*-act as a magical rite performed on these objects for the purpose of "transforming the symbol

202 E VI 184,11–12.

203 E VI 177,14.

204 See p. 36 n. 85. Note, too, that the *pcj*-land emerges from the water, thus separating from the chaos-element, on the command of the *šbtjw* (E VI 182,10).

205 E IV 358,9.

206 Chaos-divinities which represent the elements of darkness and water and take part in the transition to cosmos, see Sauneron and Yoyotte 1959 p. 52 ff.

207 E IV 14,8.

208 Reymond 1962b p. 44.

into its real shape, here, for instance, the *jḫt* of the earth into land called *pcj*"[209].

A correlation like this between the cosmogony texts and the foundation texts seems both logical and natural, even if a more precise definition of the *sw jḫt t3* might be difficult to prove. But as we have seen, the consecration of the temple ground includes the deposition of objects in foundation pits – plaques of precious metals, amulets, models of instruments, etc.[210], ingredients needed for the construction of the temple. The *jḫt t3* of the mythical *pcj*-land might have a ritual parallel in these objects deposited in the actual foundation site, the mythical preparation of the *pcj*-land for the founding of the temple having a cultic counterpart in the ritual preparation of it in the form of foundation offerings in the ground for the production of the temple – the ritual conception of the fertile earth ready to yield its crops. Seen against this background, the temple would be conceived of as growing from the earth, representing its produce.

The *sw*-act of the *šbtjw* may in the foundation context have consisted in the uttering of formulas, even if the act may have been differently conceived of in the cosmogonic context.

K3 is also mentioned in the foundation texts – and apparently is synonymous with *god.* The word *k3* denotes the god as a complex of manifested life-capacity, but in this chapter we shall not go deeper into the nature of the cosmological aspect of *k3,* only focus on one meaning relevant for the immedate interpretation of the texts: The ka represents god as the *divine majesty,* it stands for his *kingship personality*[211].

In the building text of the naos and in that of the enclosure wall, it says that the ka performed the praisegiving[212], an act which is – as will be shown – a calling into existence. This function associates with a particular cosmogony text according to which the ka utters words to the effect that the temples are established: *The words passed, uttered by the ka before being seen, this falcon, the ka who founds the temples (ḏ3jsw n snj m ḏdt n k3 n m33tw.f bjk pn k3 smn gs-prw)*[213].

The creative word of the ka can also be in the form of an order, as in the long cosmogony text which says that the ka has *decreed* the building of the temple[214]. Here, then, the ka represents the will of Horus, and the relevant

209 *Op. cit.*
210 See p. 57.
211 See Gardiner, in PSBA 38 (1916), p. 50, and in JEA 36 (1950) p. 7 n. 2.
212 E IV 14,8.
213 E VI 17,6–7.
214 E VI 185,2.

sentence might even be paraphrased: *The place is built for you as the place which you have decreed (ḫnm bw ḫr.k m bw wḏ n.k k3)*[215].

In both cases, the ka is manifest in words effecting the coming into being of the temple.

Another trait to be noted is that the divine ka of Horus can be identified with the ka of the *king*: the kingly office is the ka of Horus[216]. Thus the building text on the enclosure wall ends with the following words:

- and they (= Horus and Hathor) made enduring his image on earth: the king of Upper and Lower Egypt, Ptolemaeus X, son of Re, Alexander I, *whose ka will be rewarded with strength and victory on the throne of Horus, in front of the living ones, forever (fk3tj k3.f m kn nḫt ḥr st Ḥr ḫntj cnḫw ḏt)*[217].

The merits attributed to the ka of the king in this quotation, are those of the powerful and victorious Horus – on whose throne the king sits.

This is an interesting function of the concept from our point of view, as it bears witness to another link between the cosmogony and the building of the temple, over which the king presides as the master builder and executor of the will of Horus. A text on the inner face of the enclosure wall is illustrative in this respect. It presents a juxtaposition of the cosmogony and the temple construction, and of the Majesty of the god and the Majesty of the king. The cosmogony opens with the coming of the god to the place where the resting place is to be founded, while the land is surrounded by the flood: *His Majesty, the great one, came to the water, after the reed had arrived at its edge (but) before the two gods stabilized the ḏb3-floater in the water (spr ḥm.f c3 ḫr wcrt ḏr jw nbjt m gs.s ḫnt ḏd nṯrwj ḏb3 m njw)*[218]. It goes on to narrate that *Ḏb3* and *Wṯst-Ḥr* came into being etc., and it ends with the following words: *And thus the Seat of Re-and-Horus came into being in this place, lasting forever. And these gods praised this seat as the horizon, which is under Re; and they praised their son whom they love, the king, Ptolemy … (ḫpr nst n Rc ḥnc Ḥr m st tn r mn ḏt ḥs nṯrw jpn st tn mj 3ḫt wn ḫr Rc ḥs.sn s3.sn mr.sn nsw …)*[219]. *His Majesty (ḥm.f)* in the opening of the cosmogony is the majesty of the god coming to the place where his resting place is to be founded. But the word connotes the Ptolemaic king, because at the end of the cosmogony the latter is the majesty praised by the gods when the resting-place has been created. The parallelism

215 *Ad* the commanding word of god, see E. Otto 1964 p. 14 f.
216 Gardiner 1950. Further Schweitzer 1956 p. 72.
217 E VII 20,3–5.
218 E VI 14,13–14.
219 E VI 15,10–11.

between the mythical and the actual majesties can hardly be overlooked. The king's role as builder is here understood in the light of the mythical conception of the temple, he performs his task in the role of the divine majesty[220]; the purpose of the cosmogonic passage is to show this meaning of the king's building operations.

3. The nomenclature of the temple connotes the cosmogony

In addition to the passage explicitly narrating the building of the temple, the cosmogony text also presents implicitly a creation of the temple of Edfu, namely, through the very bestowal of the place-names. They are found again in the temple nomenclature where they designate the temple as a whole or in parts. The place-names of the cosmogony thus have three layers of reference: to the mythical topos, to the geography of Edfu, and to the temple of Edfu. In this manner the creation of cosmos is echoed by the temple-names: the names bestowed are not only names of Edfu and its localities, but names of the temple and its localities as well.

The building texts in particular make it clear that there is a nominal correspondence between the mythical-geographical cosmos and the temple building. The great building texts of the temple are engraved in broad bands on the exterior of the naos[221], in the fore-court[222], on the inner face of the enclosure wall[223], on the outer face of the enclosure wall[224], at the main entrance of the temple and on the doorway of the temenos[225]. Information regarding the temple names can also be extracted from the foundation texts[226].

Survey of temple-names having cosmological connotations

Ḏb3: In the building text of the naos *Ḏb3* is name of the temple[227], through a wordplay on *ḏb3: retribute, punish*[228] and an allusion to the punishing of the enemy by the god.

220 As regards this role of the king, see Hornung 1966 p. 26 f.
221 E IV 1,13–20,4; translated by de Wit, CdE 71 (1961).
222 E V 1,10–10,16.
223 E VI 5,5–18,15.
224 E VII 1,9–27,15; translated by de Wit, CdE 72 (1961).
225 E VIII 152,12–17; 159,5–9; 160,7–11; 161,6–12; 162,16–163,2; 164,9–12.
226 E II 31; III 105, 114; IV 352; VI 168; VII 49.
227 E IV 10,9.
228 Wb V 556,8.

Wṯst, Wṯst-Ḥr: a) The name denotes a place within the temple – a chapel situated on the eastern side of the ambulatory[229].

b) The name denotes the temple as such[230] in the building text of the enclosure wall[231], of the naos[232], and in the texts on the *St-wrt*-sanctuary[233].

D3t-n-b3: a) The name denotes the so-called "crypts" of the temple, two chapels lying to the west of the *Msn*-chapel[234]. They are especially connected with the cult of Osiris.

b) The name denotes the temple[235], or more precisely, the interior of the temple[236], which, through this designation, acquires the meaning of the dark pre-cosmos whence the god comes forth as *Cpj.*

Ḥcjt-wrt denotes the temple in the building text of the naos[237].

Ḥcjt-n-3ḫtj: Mound-of-He-of-the-horizon, is a temple-name in the building text of the enclosure wall[238].

Bw3 k3jt occurs as temple-name in a foundation text on the inner face of the enclosure wall[239].

Bw-bḥn-ḫftjw and *Ḥtm-ḫftjw* are not documented as temple-names as such; but similar names of the same content are, for instance, the name *St-sḫr-Sftḫ: Place-of-destroying-the-evildoer,* which is in the building text of the enclosure wall given as a name to the forecourt: the forecourt is referred to as *Place-of-destroying-the-evildoer, the enemy of He-of-the-horizon (St-sḫr-sftḫ ḫftj n 3ḫtj)*[240]. The name has associations with a cosmogonical record on the inner face of the enclosure wall, according to which Horus is presented as the

229 E I 283,4 f.; cf. Alliot 1949 p. 124 f.

230 It is a curious trait that several of the names have references both to a chapel in the temple and to the temple as such. A similar phenomenon can be observed at Denderah, and Gutbub gives the following suggestions to an explanation: 1) A chapel has been specialized for a general function of the temple, or 2) a chapel-name has become a designation of the temple as such, on grounds of its importance, BdE 32 (1961) p. 308 f.

231 E VII 6,1.

232 E IV 11,6.

233 E I 13, right; 18,34.

234 Chassinat room F; Porter and Moss room XII; E IV 5,5; E VII 13,6; and Chassinat room G; Porter and Moss room XIII; E IV 5,4; 13,11; E VII 13,3. – Brugsch DG p. 887; Gauthier DG VI p. 88. For the references to Porter and Moss: cf. plate 9.

235 E V 29,10; Brugsch DG p. 887.

236 E V 29,10; Alliot 1944 p. 500 n. 6.

237 E IV 4,6; Brugsch DG p. 560; Gauthier DG IV p. 166.

238 E VII 2,3.

239 E VI 321,5.

240 E VII 18,9.

protector *whose spear overthrows the enemy (mcb3.f ḥr sḫr ḫftjw)*[241], and also to the cosmogony on the naos, which towards its end reports the founding of *Place-of-punishing-the-evil-doer (St-ḏb-sftḫ)*[242].

Mspr-n-m3t is a temple-name[243].

Msn: a) The name denotes the chapel lying at the extreme end of the temple building, axially north of the chapel of *St-wrt*[244]. It is called the "first" chapel and the great throne of the Falcon *He-of-the-mottled-plumage (s3b šwt),* in the building text of the enclosure wall[245]. The chapel is said to contain a naos of black granite, within which are a statue of Horus in the falcon form and a statue of Hathor. Close to the naos is another Horus-statue, that of the *Falcon-of-gold (Bjk-n-nbjt).* In the building text of the naos the same information is given: in *Msn* the image of *Horus-of-the-horizon (Ḥr-3ḫtj),* rests in the form of *He-of-the-mottled-plumage,* together with Hathor and the *Falcon-of-gold*[246]. *He-of-the-mottled-plumage* is also called the *Sj3*-falcon[247]. In this chapel the two sacred spears of Horus are set up[248].

The chapel is also called *Seat of the Sj3-falcon*[249]. The *Sj3*-falcon is represented in his form of *gmḥsw,* i. e. a squatting falcon, holding in his hand the *nḫ3ḫ3*-flagellum[250]. When the *Sj3*-falcon in the long cosmogony is said to settle together with the image of The One-great-of-strength, looking at *(gmḥ),* Horus-of-the-*sḫm-ḥr*-weapon[251], this location in the temple is probably thought of.

b) The name denotes the temple in the building text of the enclosure wall[252].

Bḥdt: a) The name denotes a place within the temple, as it is the name of the chapel situated on the eastern side of the ambulatory[253]. It contains an image of *Mḥt* and the ennead keeping watch over Osiris[254].

241 E VI 15,5.
242 E IV 359,2.
243 Gauthier DG III p. 60.
244 Chassinat room I; Porter and Moss room XV, cf. plate 9.
245 E VII 13,1.
246 E IV 13,7.
247 E IV 5,2.
248 Alliot 1949 pp. 314–325.
249 E IV 5,2.
250 Wb II 306.
251 E VI 183,7.
252 E VI 9,8; VII 3,4.
253 Chassinat room M; Porter and Moss room XIX, cf. plate 9.
254 E VII 14,5.

b) The name denotes the temple in the building text of the naos[255]. Also in the building text of the enclosure wall[256], and in a foundation text of the enclosure wall, inner face[257], the name refers to the temple.

St-wrt: a) The name denotes a place within the temple, being the name of the chapel lying in the core of the building[258], around which the other chapels are located. The chapel contains the boat of *He-of-the-mottled-plumage,* according to the building text of the enclosure wall[259], and the god himself rests in his grand naos of black granite at the side of the boat. – Also the building text of the naos refers to the chapel under the name of *St-wrt*[260].

b) The name denotes the temple in the building texts of the naos[261], and of the surrounding wall[262].

Nst occurs as temple name in two forms: *Nst-Rc: Throne-of-Re* and *Nst-ntrw: Throne-of-the-gods.*

Nst-Rc: a) The name denotes the chapel lying south of the chapel of *Mansion-of-the-leg* and its annexe[263], and is the place where Re rests[264].

b) The name denotes the temple in a text of the southern temenos-doorway[265].

Nst-ntrw: a) The name denotes the chapel situated south of the Crypt on the western side of the ambulatory[266], according to the building text of the naos[267] and of the enclosure wall[268].

b) The name denotes the temple in the text of the southern temenos-doorway[269].

Bw wr: In the building text of the naos the name denotes the temple[270].

255 E IV 5,8.
256 E VII 9,10.
257 E VI 321,1. 2.
258 Chassinat room A.
259 E VII 15,3.
260 E IV 5,9–10; 12,9; 13,12.
261 E IV 2,9.
262 E VII 1,12.
263 Chassinat room L; Porter and Moss room XVIII, cf. plate 9; E IV 5,8; E VII 14,4.
264 Alliot 1949 p. 125 ff.
265 E VIII 161,10.
266 Chassinat room E; Porter and Moss room XI, cf. plate 9.
267 E IV 5,6.
268 E VII 14,2.
269 E VIII 161,10.
270 E IV 7,5.

The Room of the Nile: Also the river, this all-dominant element in the cosmogony, has its chapel in the temple. It lies to the west of the 2. hypostyle hall[271] and has its own door leading to the great ambulatory around the naos, through which the daily libations were brought. The sanctuary is consecrated to the The Assembly that governs the inundation, to which belong Nun, Hapj, Thoth and Ptah[272].

Thus, the nomenclature of the temple echoes the place-names of cosmos.

III. The ritual aspect of the cosmogony

a) *The mode of creation*

Perhaps the clearest evidence of the connexion between the cosmogony and the temple-cult lies in the literary form of the cosmogony myth – a form which is conditioned by a certain mode of creation: the cosmogony is effected through recitals of long series of designations, by which cosmos is called forth and appears in agreement with the content of the designations. These recitals performed by gods are basically a ritual way of creating, and they clearly fit into a cultic context. – In effect there are two series of designations activated in the cosmogony, 1: designations of the mytho-topographical landscape, and 2: designations of the geographical area of Edfu. They are homophonous and also identified with each other, but their respective frames of reference are well distinguished – they constitute two separate series.

The recurring word for creation is in the text *ḫpr*[273]. It is used to denote the coming-into-being of the actual town of Edfu. This coming-into-being is brought about by a name-giving in which the designations of the mythical landscape are said to *become, ḫpr* names of the town. Thereby an identification between the mytho-topographical place and the geographical place has been effected. The creative procedure is presented in two stages: 1) A god utters the designations of the landscape, and 2) the designations are changed into names of the town. The formula applied to denote this final phase of the creation is *ḫpr... m rn njwt tn,* which can be translated as ... *became name of this town,* or ... *came into being as name of this town.* Either translation

271 Chassinat room A'; Porter and Moss room I, cf. plate 9.
272 Bonneau 1964 p. 379; E II 255 f., I pl. XLIV.
273 Wb III 260.

ascribes to the name-giving a creative capacity; the mytho-topographical place is thereby transformed into the actual town of Edfu and its vicinity, *ḫpr* being used to denote creation by metamorphosis. The actual, material coming-into-being of a precisely located geographical cosmos is effected through this *ḫpr*-creation.

Thus, the formula *became name of this town* implies the coming into existence of the town of Edfu, the town being created in conformity with the mytho-topographical content of the names bestowed. All namegivings are combined with wordplays, by which the connexion between the mytho-topographical place and the geographical place is established – the similarity between the designation and the name expressing an identity. A survey of the names bestowed will show the technique applied. They are presented according to the order in which they are presented by the text. The kind of text we are dealing with is thematically a cosmogony, but formally – in major parts – an enumeration of designations and names, justifying the title of the book, *Enumeration of the mounds of the first time.*

Not all of the topographical designations, however, refer to mounds. Thus the list opens with the designations of the floater of reed in the water:

floater: ḏb3 and
support of Horus: wṯst Ḥr – become name(s) of this town:
ḫpr Ḏb3,
ḫpr Wṯst-Ḥr
m rn njwt tn (182,3).

Next follows

the underworld of the ba: d3t n b3 – which becomes name of this town:
ḫpr D3t-n-b3 m rn njwt tn (182,6).

The following names are connected with wordplays upon the name of a god or a divine event. The arrival of

The Beautiful-one: Nfr, and
Horus-of-praisegiving: Ḥr-j3w – has as its consequence that the names of *Beautiful Island* and *Ḥr-j3wt* are given to this town:
ḫpr Jw-nfr,
ḫpr Ḥr-j3wt
m rn njwt tn (182,11).

Similarly, the names of

Ḫnm-jtn, and
Tpj-t3wj (182,13) are town names based upon wordplays associated with a divine event: *The sun unites with the heaven upon The Two Lands: ḫnm jtn nnt tp T3wj.*

Then, to the question *What has arrived at this place?* the following answers are given:

a high mound: ḫcjt wrt,
a place of expelling the enemy: bw bḥn ḫftjw
new mspr-land: mspr n m3wt,
land of The Great-of-strength: t3 Wr-ḫpš.

The designations become names of this town:

ḫpr Ḫcjt-wrt,
ḫpr Bw-bḥn-ḫftjw,
ḫpr Mspr-n-m3wt,
ḫpr T3-Wr-ḫpš,
m rn njwt tn (182,15).

By the invocation made by *The One Great-of-strength (Ḫpšj),* another series of mounds is called forth:

mound of the radiant one: j3t j3ḫt
island of Re: jw Rc,
that which establishes the earth: ḏd t3,
high mound: bw3 ḳ3jt,
b3ḳt-land: b3ḳt,
that which is powerful of ka: wsr k3,
⸢born-of-water⸣: ms-nt,
prosperous of seats: wḏ3 swt.

The designations in turn become names of this town:

ḫpr J3t-j3ḫt,
ḫpr Jw-Rc,
ḫpr Ḏd-t3,
ḫpr Bw3-ḳ3jt,
ḫpr B3ḳt,
ḫpr Wsr-k3,
ḫpr Msnt,
ḫpr Wḏ3-st
m rn njwt tn (183,14–15).

The series of names ends with the conclusion: *The seat: bḥdt* is there until eternity, *existing upon the mounds: wnn tp j3wt* (183,15), a statement that expresses the cosmological function of the mounds and gives occasion to another wordplay:

ḫpr Bḥdt,
ḫpr Tpj-j3wt
m rn njwt tn (183,15–16).

The following puns are associated with an action of the god:

> *"May you be powerful! May you be powerful! O, Horus! as the terror which is behind the high seat (st wrt) that is there, and which destroys the enemy through its strength (ḥtm ḫftjw m nḫt.s)!"* (183,16–17); and the following names are given to *this town:*
>
> *ḫpr St-wrt,*
> *ḫpr Ḥtm-ḫftjw*
> *m rn njwt tn* (183,17–18).

Added to these appellations there is one denoting *seat* which is associated with a wordplay on *nt s3tw: the flood of the earth* and given as the name of *Nst: The Throne* to *this town:*

> *ḫpr Nst*
> *m rn njwt tn* (183,18).

Next comes a praisegiving which introduces new places in the mythical landscape, the designations of which are turned into names of *this town:*

> *Giving the praise (ḥsj-Rc):*
> *high mound: ḫcjt wrt,*
> *mspr-land ⸢of the circumference⸣: mspr ⸢šnw⸣,*
> *high place: bw wr –*
>
> *ḫpr Ḫcjt-wrt,*
> *ḫpr Mspr⸢-šnw⸣,*
> *ḫpr Bw-wr*
> *m rn njwt tn* (184,1–2).

The designation

> *place of the throne: bw ḥmr,* is uttered by *C3*

and becomes name of *this town:*

> *ḫpr Bw-ḥmr*
> *m rn njwt tn* (184,3).

In most of the instances noted the words are *uttered*; the creative word is a word recited. This is expressly indicated by adding *so said: j.n* (the god); or the words belong to an *invocation: njs.* The creation is a calling into being.

Behind this way of creating there lies a particular conception of the word: the word partakes in that which it denotes[274] in such a way that its content is

274 See Sauneron 1966.

released when it is uttered[275]. This belongs within a ritual context, the creative recital being a typically ritual way of creating.

The creative capacity of the word can be conceived mythologically. There are different myths about the creative word in Egyptian literature. Also the *ḏ3jsw*-words have their myth. It is not told in our text or in other Edfu texts, although it is implied by our cosmogony. In the Esna-material, however, we find it recorded. The myth of the *ḏ3jsw* is engraved on one of the pillars in the pronaos of the temple of Esna[276]. It expresses metaphorically the import of the creative word. The *ḏ3jsw*-words are said to go forth from the mouth of *The Heavy Flood (Mḥt-wrt)* which in the traditions of Esna is personified and identified with Neith, the creator[277]. Their leaving her mouth inaugurates the creation of cosmos. Seven *ḏ3jsw*-words are uttered by the goddess, and they leave her mouth personified into seven creative divinities – the *ḏ3jsw*-commanders. Their being uttered is analogous to a birth from the mother goddess.

Also in the Edfu cosmogony the *ḏ3jsw*-words belong to The Heavy Flood[278]. They are the enumerations of the mounds. The words are uttered by the creators and their content materializes. In the Edfu cosmogony they are said to *go forth (snj)*[279] (i. e. from the mouth of the creator); and as they go forth, their content manifests itself: it is *seen (ḫf)*. Through the recitals of the designations of the mounds, the mounds appear. A. Gutbub makes allowance for this conception when he translates the title of the book: *Annoncer (le nom) des buttes de l'époque primordiale*[280], a translation which indicates a functional purpose of the book: it is not to be silently read by a devoted priest, but to be recited for a creative purpose.

It should be stressed that the enumeration is not a catalogue for the sake of survey, but a catalogue for the sake of creation. The words are the words of the creator; when recited they call into being a cosmic landscape.

Similarly, when the landscape called forth through the *ḏ3jsw*-words is turned into the geographical place of Edfu, this is done by way of enumerating names; the creation of Edfu is presented in the form of a series of glosses. It is important for the understanding of the creative procedure adopted by the text to bear in mind that these glosses are not to be regarded as mere interpolations, disturbing the flow of the creation story, but that they are an integral element

275 Morenz 1975.
276 Sauneron, *Esna* III (1968) No. 296 § 18.
277 Sauneron, *Esna* V (1962) p. 268 f.
278 E VI 181,11.
279 E VI 182,9; 11–12.
280 Gutbub, *Kom Ombo* I (1973) p. 126.

of the creation story[281]. They can be said to represent a special exegesis of the coming into being of the mythical landscape, but they are nevertheless themselves integrated into the cosmogonic process and have become part of it. The names of Edfu are identified with the *ḏ3jsw*-words of The Heavy Flood; and through this identification Edfu comes into being; the glosses cannot be left out without serious damage to the cosmogony. From the point of view of textual analysis, the relationship between the two name-lists can be said to be one between a mythical name-list and an existential exegesis of it. To both kinds are, though, ascribed specific creative functions in the cosmogony, as the first-mentioned calls into being the cosmic topography, while the latter calls into being its geographical location.

An adequate interpretation of these cosmogonic name-lists cannot be achieved unless they are seen in their functional context. They are meant to be recited by the creators; they are invocations. *The words of The Heavy Flood (ḏ3jsw njw Mḥt-wrt)* invoke cosmos; this is demonstrated by the mode of creation presented by the text.

There is one contextual characteristic of the enumerations and name-givings which should be noted, because it offers additional testimony to the ritual character of the cosmogony, and lends to it an undoubtedly *cultic* aspect: The recitals are – and, as a rule, are explicitly said to be – performed within the context of praisegiving. Thus we find that the designation of *the underworld of the ba* is uttered in a praisegiving context:

... "praise!" so said the four. "Look!" so said the two gods, and so said the šbtjw. "Who has come < out of > the underworld?" so said the šbtjw. "The underworld of the ba is this place!" so said the Falcon. And the Underworld-of-the-ba became name of this town (j3w j.n fdw m j.n nṯrwj j.n šbtjw ptr < m > d3t j.n šbtjw d3t n b3 bw pn j.n Bjk ḫpr D3t-n-b3 m rn njwt tn) (182,5–6).

Similarly, praise is given for entering the place of building (183,10); the consequent naming of the mounds of the *pcj*-land, on which the building is going to take place, is also performed within a praisegiving context: *The One-great-of-strength made the invocation, praising the pcj-land: "Mound of the radiant one!"* etc. *(njs Ḫpšj m ḥsj-Rc pcj: j3t j3ḫt* etc.) (183,12).

Another *praisegiving (ḥsj-Rc)* follows, which introduces another series of mounds of the *pcj*-land: *"High mound!* etc. *(ḫcjt wrt,* etc.) (183,18).

A praisegiving *(ḳcḥ j3w)* is also performed for the protection of the place (184,4).

281 This view differs from that of Alliot, cf. Barucq's commentary on his translation, 1966 p. 152 ff.

The praisegiving is the cultic form of the name-recital. Egyptian hymns and laudations are a combination of praisegiving and enumeration of appellations and names, the function of which is to evoke the divine presence or the divine qualities denoted by the names recited. Praisegiving and hymns actually look like catalogues; the typical Egyptian hymn is a list of names and epithets. But they are not lists for the sake of survey. They are meant to be recited, for the recital is an evocation. Any praisegiving, or hymn, whether to the god or to his place, has a creative potential released during the recital.

The divinities in our cosmogony create in cultic-ritual fashion. Not unexpectedly this corresponds to a cultic-ritual namegiving of the temple, which in characteristic points parallels the namegiving of cosmos.

b) *The ritual bestowal of names on the temple*

A ritual bestowal of names on the temple is documented. It is presented as being performed by gods during the *ḥts*-festival, i. e. the festival of completing the temple and handing it over to its divine lord[282]. It is likely that it refers to an actual rite enacted on this occasion. Whether or not – its cultic-ritual nature is marked, the namegiving is presented *qua* rite.

The *ḥts*-festival is documented in pictures[283] and texts[284] in the upper register on the inner face of the enclosure wall. In the centre of the scene the triad of Horus, Hathor and Ptah is seated in a naos. Before them, also within the naos, stands the king, holding two *ḥts*-signs in his hands[285]. Behind him and outside the naos are Re-Harakhte and Tanen, lifting their hands in adoration, and behind them are Thoth and Seshat with writing tablets in their hands[286]. Behind these four gods are two rows of creators[287]. They include *C3* and *W3*.

Two of the texts belonging to the scene are especially interesting. One is a recital by Re-Harakhte, the other by Tanen. The recital performed by Re-Harakhte is in the form of a praisegiving directed to the High Seat: *Words to be uttered by Re-Harakhte, great god, lord of heaven, He-of-the-mottled-plumage, who comes out of the horizon praising the High Seat with all its beautiful names that exalt its might among the nomes (ḏd mdw jn Rc Ḥr-3ḫtj*

282 Wb III 203; 303,1.
283 E X pl. CLII.
284 E VI 318 ff.
285 Wb III 202,7. 8. Gardiner, Eg. Gramm. signs Aa 30,31.
286 E XIV pl. DCIV.
287 E XIV pls. DCIV–DCV.

nṯr ͨ3 nb pt s3b šwtj pr m 3ḫt m dw3 St-wrt m rnw.s nfrw sḳ3 b3w.s m sp3wt)[288].

Then follows the praisegiving, in which the place is given its characteristic appellation: *(O, you) seat in which Apep has been pierced! (st wnp ͨpp).*

Thus having designated it as a place of victory, he goes on to say that he has made it a *house-of-appearing (pr-ḫ ͨj)*[289] in *The Mansion-of-the-Falcon*[290]. He has *hidden (jmn)* himself in it; and he has made his *great place (bw wr)* in the *first shrine (sḫm ḫntj).* He has *made its glories,* he has *beautified its seats,* he has ⌜...⌝ *its chapels,* the *Ḥtp-nbwj*[291], the *ͨḥ-m-ḥrt,* the *Ḥt-nḥḥ* and the *Pr-ḏt.*

The consequence of the description is that the places corresponding to it are said to *come into being (ḫpr):*

And The Place-of-piercing came into being,
and The House-of-appearing came into being,
and The Mansion-of-the-seat came into being,
and The Place-of-hiding came into being,
and The Great Place came into being,
and The First Shrine came into being,
and The Making-the-glories came into being,
and The Beautiful-of-seats came into being,
and ⌜...⌝-of-chapels came into being,
and Ḥtp-nbwj came into being,
and ͨḥ-m-ḥrt came into being,
And Ḥt-nḥḥ came into being,
and Pr-ḏt of gold came into being,
in this temple

(ḫpr St-wnp,
ḫpr Pr-ḫ ͨj,
ḫpr Ḥt-Bḥdt,
ḫpr St-jmnt,
ḫpr Bw-wr,
ḫpr Sḫm-ḫntj,
ḫpr Jrj-3ḫw,
ḫpr Nfr-swt,
ḫpr ⌜...⌝-sḥw,

288 E VI 319,3–4.
289 *Ḫ ͨj* refers to the cultic appearance in the temple, Wb III 240.
290 Official name of the temple.
291 The name associates to the mythical *ͨ3* and *W3, united in peace.*

ḫpr Ḥtp-nbwj,
ḫpr Cḥ-m-ḥrt,
ḫpr Ḥt-nḥḥ,
ḫpr Pr-ḏt m nb
n ḥt-nṯr tn)[292].

Tanen makes a similar recital. He says that he has made a *house-of-appearing (pr-ḫcj)* in *The Mansion-of-the-seat*[293]. He goes on describing it in the same way as did Re-Harakhte, and designations applied become names of the temple:

And The Place-of-piercing
and The House-of-appearing
and The Mansion-of-the-seat
and The Place-of-hiding
and The Great Place
and The First Shrine
and The Making-the-glories
and The Beautiful-of-seats
and ⌈...⌉-of-chapels
and Ḥtp-nbwj
and Cḥ-m-ḥrt
and Ḥt-nḥḥ
and Pr-ḏt
became name(s) of the temple (k3 m ḥt-nṯr)[294].

The two enumerations are identical, but for one trait: The first recital presents a coming into being of (places within) the temple, while the second presents a coming into being of names of (places within) the temple. The difference is not essential. Both enumerations imply a coming into being of the temple; the first leaves out the concluding statement that it is a ritual creation through namegiving, which is unnecessary in this case, as the information is given in the introductory words: *Words to be uttered by Re-Harakhte ... who comes out of the horizon praising the High Seat with all its beautiful names ...* The case is, though, a clear indication that the namegiving is conceived of as a coming into being of the content of the name.

It appears that the recitals of names performed by the two gods are of the same kind as the recitals made by the creators in the cosmogony with regard to the following features: 1) In both contexts they conform to the same

292 E VI 319,6–8.
293 Official name of the temple.
294 E VI 321,2–4.

pattern of procedure. A god designates the place with reference to its cosmological functions or functions as a sacred place, and the designations then become names – of cosmos or the temple – through wordplays. 2) The names are uttered in a praisegiving context: The sanctuary is *praised* with its names – they *exalt* its might. The creative effect of the praisegiving is no accidental circumstance but its very functional *raison d'être*[295]. As has been pointed out, its purpose is to evoke the divine qualities indicated by the names. – In a similar way the *pcj*-land is named in a praisegiving. The cosmogony is like a hymn to the world, recited by the creators, a *magnificat* to the world of god.

The cultic context of the namegivings of Re-Harakhte and Tanen is the inauguration of the temple. The building has been erected and is now undergoing a sanctification in which it is ritually transformed according to the names bestowed upon it. After this event it is ready to be delivered to its divine lord, so that he may enter it: *Darkness (Kkw)*, a member of the Ogdoad, exclaims that the assembly of creators praises the god on *your beautiful festival of entering your house (m ḥb.k nfr n cḳ pr.k)*[296]. The role given to *Darkness* is here conditioned by the time of the ritual creation: the moment when Re-Harakhte *comes out of the horizon (praising the High Seat with all its beautiful names)*, in other words, the very same moment when *cosmos* is created, according to the cosmogony myth.

Ptah makes the exhortation to the god: *Re, enter your sanctuary, that I may enter after you! (Rc cḳ c3jt.k cḳ.j m-ḫt.k)*[297].

The summon has its parallel in the concluding words of the cosmogony: *Come, you who are our lord, Harakhte, image of Re! The place is built for you as the place decreed for you by the ka!*[298]

The timing of the ritual event, and the presence of the Ogdoad, Tanen, Thoth, Seshat, Ptah and the rows of collective creators record a cosmogonic undertone in the hymnic creations. It becomes even more audible in the words uttered by *W3j: We praise you! We stabilize your floater and it uplifts your Majesty in Wṯs-Ḥr! (dw3.n n.k ḏd.n ḏb3.k wṯs.f ḥm.k m Wṯs-Ḥr)*[299]. With these words pronounced by the one who stabilized the mythical *ḏb3*-floater, the two parallel creations of cosmos and of the temple are brought together. The construction of the temple is seen in terms of the mythical act of stabilizing the floater that uplifts the god above the water of the river.

295 See Assmann 1975a.
296 E VI 321,17.
297 E VI 321,16.
298 E VI 185,2.
299 E VI 323,4.

IV. Conclusion: The temple reveals the sacred nature of the world of man

The parallelism demonstrated between cosmos and the temple shows that the Egyptians developed the cosmological aspect of the temple intentionally and coherently. The temple witnesses a religious conception of the world; it expresses it in a variety of ways and unfolds it as systematically as any theological treatise. This monumental edifice is the Egyptian exposition of the sacred nature of cosmos.

The dwelling-imagery, in so far as it displays the world of man in its cultural and economic activities,' is drawn into this cosmological aspect: the temple represents the world of man not only in the sense of Egyptian *nature,* but also in the sense of Egyptian *culture.*

The temple is a religious exegesis of the world of man in all its aspects, though with one limitation: the world of man is the world of Egyptian man. We are studying a concept of cosmos which excludes that which is outside Egypt. In the cosmogony texts, the geographical borderline is drawn even more exclusively, coinciding, as it does, with that of Edfu.

Edfu, with its urban and agricultural area, is conceived of as the sacred place of Horus, and understood with reference to this god: It is the emergent mound of the divine creator, it is his place of victory, it is his cultic seat – the place where he is adored and it is his home. It is only to be expected that designations like The Mansion-of-the-Falcon and The Mansion-of-Horus are also given as names to the *town,* as in the list of town names recorded in the forecourt which opens with the names *St-wrt, Ḏb3* and *Wṯst-Ḥr* and goes on with *Ḥt-Bjk: The Mansion-of-the-Falcon, Ḥwt-Bjk: The Chapels-of-the-Falcon* and *Ḥt-Ḥr-Ḥrw*[300]: *The Mansion-of-Horus-of-the-Horuses* (E V 396,1–2).

As the material form of a religious interpretation of the world of man, the temple has the function of a cosmophany: it reveals the sacred nature of the world, in that it documents it as a world created by god to be his abode, and his place of adoration, i. e. where he is magnified.

This relationship between god and the world is a close one, and it finds expressions that go deeper than do the images of the master and his house, and of the creator and his work of creation.

300 As regards the reading *Ḥt-Ḥr-Ḥrw,* see Gutbub 1953. Other examples in E VI 205,5: *Ḥt-Bjk,* and E VI 206,7: *Ḥt-Ḥr.*

B. The temple is a theophany

I. The mythological background of the theophanic character of the temple

The temple reveals the nature of the gods. It shows their cosmic character and their cosmic origin – through its architecture and decoration and by ritual enactments of their theophany which is staged as the twofold appearance of cosmos-and-gods. The theological premises of this function of the temple are expressed mythologically in the cosmogony text. The cosmogony is not only about the coming of the world but also about the coming of the gods.

a) *The mythical theophany*

The cosmogony myth presents an interpretation of the world of man which ascribes to the latter a meaning of theophany: the places that appear never do so in their own right but always together with a god. The events of the annual inundation and the daily rising of the sun and the emerging banks and mounds are seen in terms of divine appearances. Also the changes which the places undergo – such as the canalization of the river and the maintenance of the canals – are connected with divine appearances. The relationship between the gods and the places is not theologically defined in the myth; but their comings are presented as simultaneous events.

The series of theophanies starts with the delimiting act performed by the two Delimitors resulting in the pacification of the torrential water, which is changed into water of order and stability with the coming of this divine pair, *united in peace*[1].

Next the *Ḥtr-ḥr*-falcon appears with the reed floating down the river (181,14).

When the Beautiful-of-harpoon arrives, a part of the reed separates from the rest and is given the function of a floater (181,16).

Then the falcon of heaven settles upon the floater and appears as the falcon who is supported (182,2) – and with this theophanic event *Wṯst-Ḥr* appears.

1 *ḥtp* connotes here both *unity* and metamorphosis. A similar example can be seen in the tomb of Nefertari, e. g. Dondelinger 1973 p. 103.

With the coming of the Divine Ruler, *Ḥḳ3 nṯrj* as the *Sj3*-falcon, the dimensions of the cosmic space are laid out: the extension of heaven is drawn with the stretching out of the wings of the Ruler-of-flying, *Ḥḳ3-ntb* (182,7); and the underworld comes into being as the place whence the divine ba comes out (182,5–6). This appearance is described as the most marvellous of the theophanies: The Divine Ruler appears as the *Sj3*-falcon, the one with a *head beautiful-of-face, the Hovering-one spreading real marvels, making the darkness m3c with his wings, (being) mailclad and (being) majestic, clothed in s3wj-gold* (182,4–5). And this glorious apparition has a glorious cosmological counterpart: When the Falcon starts flying, the Winged Disk, *Cpj* comes into being, and in the radiance of this theophany the contents of the cosmic space appear. The sun god has revealed himself as creator. The coming of the Winged Disk – Horus in his combined solar and falcon aspects – is the cosmogonic event *par excellence.* It is a theophany which permeates the whole topos so that the contents of the dark acquire form in the light of the god.

The first cosmic phenomena to appear – from the night as well as from the water – are the mounds, lying as islands in the water. The emergence of the mounds is combined with new theophanies: Beautiful Island *(Jw nfrt)* and Horus-of-praisegiving *(Ḥr-j3wt)* come into being with the coming of the Beautiful-one *(Nfr)* and Horus-of-praisegiving *(Ḥr-j3w)* (182,10–11).

When the sun unites with the sky, the area underneath is illuminated and comes into being as The-sun-has-united *(Ḫnm-jtn)* and That-which-is-upon-the-Two-Lands *(Tpj-t3wj)* (182,14). In this area the following mounds appear: The High Mound *(Ḥcjt-wrt)*, the Place-of-expelling-the-enemy *(Bw-bḥn-ḫftjw)* and the Land-of-He-who-is-great-of-strength *(T3 Wr-ḫpš)* (182,15–16). The *mound* is closely connected with the solar creator in Egyptian cosmological traditions; in this feature, the Edfu traditions are based on an ancient mythical construction[2].

Upon the coming of the Protector, *(Nṯr-ḥn)*, who is also the Lord-of-fear *(Nb-nrw)*, the water of *p3 Ḫnw* is opened. And with the coming of the Lord-of-opening *(Nb-ptḥ)* who is the Sacred Heart *(Nṯrj)*, the canals of *Ptḥ* and *Nṯrj* come into being (183,1–2).

When the earthgod Tanen emerges *(bs)* from the water and arrives at the reed (183,3), this means that the water has subsided so as to lay bare the earth at the place where the reed is, so that it is no longer in the water. The earth rising out of the water is the main theophany of Tanen. That he is also thought to be the deity behind the opener of the canal water (183,2) is an indication that the canals are seen as integrated elements of the kind of earth he

2 See de Buck 1922 p. 23 ff., 35 ff.

manifests. The earth of Tanen is not only the mounds upon which the settlements are founded, but also – and primarily – the irrigated earth fit for cultivation.

The god Ptah also makes his presence felt in the last-mentioned cosmological events. In our text the god who opens *(ptḥ)* the water is surely identified with Ptah *(Ptḥ)*. The opening of the water is a theophany of this Lord of the inundation. He is a god often fused with Tanen[3]. In the Ptah-Tanen-constellation he appears with the irrigated land.

Tanen is also associated with the Protector. He is the maker of the latter's weapon and he is the one who brings it to the Protector: *Sgmḥ* and *Ḥr-sẖm-ḥr* are names of the weapon which is ⌈*separated*⌉ *by Tanen* (183,8). The weapon is identified as the *šm-wḏ3*-spear which the protector god carries in his hands (182,17; 183,5). At a later stage of the cosmogony it is also identified with the grand *ḥḏ*-weapon, the *ḥḏ wr,* of the Maker-of-the-earth: *Jr-t3,* a serpent god who is sometimes identified with Tanen[4] but who seems to be distinguished from Tanen in our text.

The completion of the *pcj*-land is connected with the coming of the *šbtjw* (183,11–12). With their arrival the last series of mounds appears (183,11 f). These are the mounds upon which the temple is built.

When the crew of Horus-of-protecting-his-father arrives, The Place-of-the-crew appears near the water (184,13).

Finally, with the coming of Horus to the *pcj*-land, the latter appears as his *support* (184,14).

There is thus a concurrence of appearances running all through the cosmogony. The emphasis is placed on the cosmic appearances – even though the theophanies are as a rule mentioned first – because of the pronounced cosmogonic perspective of the text which deals first and foremost with the creation of the cosmos of Edfu. The theophanic character of this creation comes as a qualification. The intention of this cosmogony is apparently not to explain how the cosmos has been created (that can be ascertained through direct observation), but to present the religious view that it is created as a world of gods.

The theological identity of the gods that appear:

Among the numerous divine actors, the *falcons* play a dominant rôle. Different falcons appear: The *Ḥtr-ḥr*-falcon is first presented – hovering above the reed

3 E I 305,4; 498,12; IV 4,2; Sandman Holmberg 1946 p. 56 ff.
4 E I 329,13; Sandman Holmberg 1946 p. 185 f.; Wb I 109.

when it comes floating to the place. As he is identified with the Protector later in the text, the warrior-aspect is attributed to *Ḥtr-ḥr*. This aspect is in agreement with the nature of the floater: it is a *protected* place, as it is tied to the harpoon. It may be added that a falcon-head is fastened to the tip of the spear in the relief to the text and that the text itself describes the spear of the protector as having the head of the *Ḥtr-ḥr*-falcon (183,6). – *Ḥtr-ḥr* is identified with the falcon who is supported by the floater, the Horus-of-the-support.

Thereupon follows the Ruler-of-flying *(Ḥḳ3-ntb)* who is identified with the Flying Wing *(Ndm ntb)*. In the relief he is depicted with his wings outstretched over the assembly of creators[5]; he is the god who stretches out *the ends of the sky*[6].

A frequently occurring falcon is the *Sj3*-falcon.

Finally the Flying Ba *(B3-ḥdd)* should be mentioned, who appears in the form of a *ba*-headed falcon in the relief to the text[7].

These are all differentiations of the falcon god of Behdet; this is shown in laudatorial Edfu texts where all falcon-names are ascribed to Horus. The relationship between the hypostases is moreover reflected in the cosmogony texts themselves. In the long version it says that the Ruler-of-flying, identified with the Flying Wing, comes to give offering to the Lord of *Ḏb3* (182,8–9); in the relief the latter is depicted as a falcon sitting on three stalks of reed. With this act of homage the celestial aspect of the falcon (the Ruler-of-flying is the falcon of the wings stretching out the ends of the sky) is transmitted to the falcon resting on his *Ḏb3*-floater, while the falcon of the *Ḏb3*-floater is given priority. The name-giving connected with this homage offered by one falcon to another reflects the connexion between them; the Lord of *Ḏb3* is magnified with the name of the Ruler-of-flying: He-of-Behdet is identified with the Ruler-of-flying (182,10). In a parallel text the intention of the homage given to the Lord of *Ḏb3* is *that he may unite with heaven*[8].

By these hypostases the falcon-nature of Horus is stressed as a central characteristic of the god.

An epigraphical indication of the dominant place occupied by the falcon-apparition, also occurs, namely in the extensive use of falcon-signs in the text to denote the deity. They differ slightly from each other and there is a relatively wide option of readings, which can cause some confusion. Theoretically the following readings are possible for all of them:

5 E XIV pl. DLX; cf. plate 1.
6 E VI 182,7; compare VI 15,1–2.
7 E XIV pl. DLXI; cf. plate 2.
8 E VI 15,2–3.

nṯr, god
bjk, falcon
nb, lord
Ḥr, Horus.

Context alone provides the clue to the reading in question. On the other hand, one should not distinguish too sharply between the signs as regards their precise denotation; most probably the meaning is in any case not exclusive but connotes the other meanings as well – incorporating god, falcon, and Horus.

The falcon-manifestations mentioned present Horus as the god of the sacred place – coming with it, protecting it, resting on it. At the same time he is the lord of the sky, the god who spreads out his wings to the ends of the world.

The *ka (k3)* has to all appearances reference to Horus as the *falcon king*. The ka is on the scene before the first seat of Horus is created. A reliable translation of the word cannot be given, but it is used synonymously with god in our text[9] and can be paraphrased as *the god*. This ka came to the reed *while this object was (still) being made to move; ⌜the Falcon was (still) in the sky⌝* (182,2). The theological identity of the ka here seems to be Horus in his falcon aspect; the god comes as the falcon hovering above the reed even before it is stabilized. This conclusion is supported by the shorter parallel text on the enclosure wall, which in the corresponding passage reads: *The ka came as the Flying Ba (pḥ k3 m B3-ḥdd)*[10]. In the long text this Flying Ba is identified with the falcon as the *Ruler-of-flying (Ḥk3-ntb)*, the name of the falcon when he stretches out the ends of the sky; and later the text identifies him with the solar falcon, the Winged Disk (see below).

The ka orders the temple to be founded: *The place is built for you as the place decreed for you by the ka (ẖnm bw ẖr.k m bw wḏ n.k k3)* (185,2). In this instance, the ka represents the god's will – the ordering Majesty[11] of the God. Also in the shorter cosmogony on the enclosure wall, the ka is the falcon ruler who orders the temple to be founded: *The words passed, uttered by the ka before being seen, this falcon (bjk pn), the ka who founds the temples (k3 smn gs-prw)*[12]. In other words, the order to found the temple has gone out before the god has appeared as the Winged Disk: he is not seen; it is, then, still night. Earlier this text has mentioned the episode, also related by the long text, of the coming out of the ka as the Flying Ba to give offering to the Lord

9 The ka listens to invocations (182,17); he speaks (184,9); he orders (185,2).
10 E VI 15,1.
11 See Gardiner, PSBA 38 (1916) p. 50 and JEA 36 (1950) p. 7 n. 2.
12 E VI 17,6–7.

of *Ḏb3* that the latter might unite with the sky, and it says that *his utterance is heard before his being seen (sḏm.tw md.f n m33t.f)*[13]. Thus the ordering ka is identical with the ka who hovers above the reed while it is still moving on the water and before it is stabilized.

As we have seen in an earlier chapter, the ka can be identified with the actual king – the double reference to the Majesty of the divine falcon and to that of the king implying that the two are identified. It can be parenthically added that also in the ritual coronation of the living falcon, bird and king are identified. The event took place at the time of the *sed*-festival – a kingship festival – and I would like to quote H. W. Fairman's account: "... there can be no doubt that it was intimately connected with the kingship. The very date is significant ..., and the texts leave no room for doubting that throughout the festival the Sacred Falcon, the king, and Horus were as one, and that the festival also celebrated the annual renewal of the coronation of the reigning king"[14].

The creator-function of Horus is most prominent with the manifestations of the *sungod*. The first mentioned of these is the Winged Disk *(Cpj)*. He is identified with Horus as the god who comes flying out from the underworld *(d3t)*, the realm of darkness, and thus the falcon-aspect is included in this solar theophany. Moreover, the Winged Disk is expressly identified with the Ruler-of-flying and the Flying Wing, Horus-forms representing the heavenly aspect of the god. Still, the predominant nature of the Winged Disk is solar. Iconographically the Winged Disk is depicted as the sundisk with wings attached to it. He is, then, the sungod – and in his first manifestation.

This outcoming sun is, according to Edfu texts, seen as the *ba* of Horus, which does not mean *soul* in a spiritual sense but refers to the concrete theophany[15]. The fact that the Winged Disk comes into being when Horus takes to flight points to another characteristic of this Horus-hypostasis: it represents his most dynamic creative aspect. Coming out of the underworld means not only coming into being but also coming as creator – the latter qualification being implied by the solar nature of the Winged Disk: the sun causes the contents of the cosmic room to appear.

Re is another name of the sungod creator. In the theophany of Re the cosmic places and their gods develop – Re being the disk that has united with the heaven of the cosmos. An especially close relationship exists between

13 E VI 15,3. Compare E VI 182,5–10.
14 Fairman 1960 p. 80.
15 See p. 135.

Re and Horus. Even if they can appear in mythological texts as individual deities, they are often seen in alliance or as a unity. When the two are united into the composite Horus Re, their respective areas constitute the entire area of the god of Behdet who appears as the creator and the protector and lord of the sacred place. Under the name of Horus Re, the god of Behdet is presented in our text when he proceeds to unite with heaven (182,12). Just as he is identified with the Winged Disk when he effects the transition from the underworld, he is identified with Re when he appears as the supreme lord of the cosmos, the sovereign creator who encompasses the dynamic Winged Disk. The god of Behdet is known – also outside the town – as Horus Re or as the two gods in alliance: Re-and-Horus *(Rc-ḥnc-Ḥr)*[16].

It may be added that Re and Horus can be identified with each other, taking over each other's attributes so that their individual features are blurred. Thus the falcon Horus is said to have a *shining face*[17]; the solar Re can appear as the falcon, as when he is greeted as *He-of-the-mottled-plumage*[18].

The *protectors* constitute a group of their own, subordinated to the creators; but they are nevertheless connected with Horus, and represent his protecting functions. The Protector, *Nṯr-ḥn,* is the image of the Shining-of-face, the *Ḥtr-ḥr*-falcon (182,17); he carries the attribute of Horus in his hands.

A clue to understanding and defining the cosmological nature of the Protector and his protective acts can be found in this attribute, the weapon. Significant features of the weapon are that it comes floating with the inundation (p. 28 note 17) and that it is elevated by Tanen and brought by this god to Horus. Tanen is the inundated earth that rises again from the water, fertilized by it. Thus the weapon intimately belongs with the inundation[19] and its life-creating powers manifested in the earth. In line with this, the weapon is instrumental in removing the stagnant water, by opening and maintaining the canals.

This life-sustaining role of the weapon is stressed by identifying it with *Wtṯ* (183,4; p. 34 note 73), and with *ṯm3-c,* or, *ṯ3j-c* (184,8). When identified with *ṯm3-c,* the stress is laid on its function to maintain the uninterrupted flow of the water in the canals, securing the prolonged effect of the creative powers (p. 45). As *He-who-is-in-front-of-the-male-one* (184,9–10) it is associated with the phallic falcon-statue in the *Msn*-chapel (p. 97).

16 Gutbub 1961 and *Kom Ombo* I 223 g, Monographie 90.

17 E VI 186,3–5; VII 11,1.

18 E IV 357,15–17.

19 Its connexion with the inundation is also prominent in the textual and pictorial representations of the Myth of Horus, where it is instrumental in Horus' war against the "water monsters" at the time of the inundation (p. 15).

It can be concluded that the functional area of the Protector is connected with the life-generating properties of the inundation – his activities intend to protect and ensure the potency of the water.

In the relief to the text, this protective relation of Horus to the water is expressed by representing the god of the temple of Edfu walking, weapon in hand, before the personified and libating canal of Edfu, *p3 Ḫnw* (E XIV pl. dlxi; cf. plate 2).

The weapon is given different names in the text, and these apparently refer to different versions of it and indicate a differentiation of Horus' protective function. Sometimes the protective activities are detached from the god and take on an independent rôle in the individual weapons. The essential connexion with the owner, however, comes to light in laudatorial acclamations and apostrophes directed to them. There is a relationship of partial and representative identification between Horus and his weapons. It is alluded to in one of the lesser cosmogony texts in a typically Egyptian way: the weapons are connected with Horus through a wordplay on *ḥr: face* and *Ḥr: Horus:*

The Protector came as Horus-of-the-mighty-face, the sgmḥ-spear, the similitude of He-of-the-shining-face as the Ḥtr-ḥr (Contracted-of-face?), the living ba for him, as the Lord-of-the-head whose face is beautiful (jw Ḥn-nṯr m Ḥr-sḫm-ḥr p3 sgmḥ m snn j3ḫw ḥr m Ḥtr-ḥr b3 cnḫ n.f m nb ḏnnt ḥr.f nfr)[20].

The weapons can be personified, though they do not seem to be so in the cosmogony. In other Edfu texts, however, they can appear as Horus-hypostases uttering sacred formulas, and their names are determined with the sign of a god[21]. Also the grand *ḥḏ*-weapon can be personified in this way[22]. And in the Mammisi, *Ḥr-sḫm-ḥr* is depicted in a relief as Horus sitting behind Hathor[23]; on his crown is the spear with a falconhead on it and a snake – the weapon emblem documenting the protective aspect of the snake[24].

The use of hypostases is a literary device frequently applied when the different divine qualities and functions are related to each other. It is a characteristic of myth, because myth is a literary category which has specialized in presenting divine qualities and functions in the form of a narrative. Sometimes the hypostasized qualities or functions are "sons" and "daughters" of a god. The cosmogony myth contains many Horus-hypostases. Where laudatorial passages are inserted, the different divine aspects personalized into

20 E VI 15,3–5.
21 E III 122,2; IV 78,6–7; 235,1.
22 E IV 78,8.
23 The Mammisi 94,16.
24 Cf. E VI 182,16: Does the text refer to this snake who is associated with the crown?

individual figures for the narrative story are assigned to one god: in laudations and hymnic apostrophes the opposite tendency can be observed – the divine qualities and functions are attributed to one god.

There are other Horus-hypostases, in addition to those mentioned above, dispersed throughout the text; some of them have names alluding to cultic functions, others have names alluding to mythological constellations. Thus there are Horus-of-praisegiving[25] and Horus-of-joy whose names have a cultic timbre; and there is Horus-of-protecting-his-father whose name alludes to one of the best known Horus-myths, namely, that of Horus as son of Osiris protecting his father from the evil Seth. Horus of the *pcj*-land is called *child of Ḥtr-ḥr,* an epithet which points to a genealogical relationship between the two Horus-hypostases: the one connected with the beginning of the creation and the introductory protective actions on the water; the latter connected with the conclusion of the creation and the protective acts taking place on the *pcj*-land. The sacred place of the first one rests on water, that of the latter rests on the mounds.

The theophany of *Tanen* is the earth that emerges from the water, and it lies behind the wordplay on his name *Ṯnn: The One-who-rises,* or *T3-ṯnn: The rising-earth*[26]. His relationship to the other gods is an interesting one. Being the fertile earth, the produce of which the gods live from (as is shown in the offering institution), he is called *father of the gods*[27]. Under this life-producing and life-sustaining perspective, he is also called the *father of fathers* and *the creator of gods*[28]. The collectives of creators including the *šbtjw* are called *children of Tanen*[29]. His paternal relationship with Horus acquires a special meaning within the mythological context of the fighting Protector, as Tanen is the producer of the protective weapon and thus is specifically seen as the source of Horus' life-sustaining capacity[30].

The God-of-the-temple appears in the cosmogony when the intention to build the temple is first formulated. He is the god who utters the words. He acts here as an individual figure although he is clearly a Tanen-hypostasis, which is ultimately identified with Horus. This is evident from the epithets given to him in the legend to his portrait in the relief. It presents him as:

25 E III 184,1.
26 See Sandman Holmberg 1946 p. 31.
27 E IV 4,2.
28 E IV 21,11; V 157,1–2.
29 E IV 353,2; 358,11; 390,4.
30 E VI 183,8.

1: *The God-of-the-temple in Wṯst-Ḥr, who came out of the water towards the High Seat (p3 nṯr n ḥt-nṯr ḫnt Wṯst-Ḥr bs m nwn r st wrt).* The words attribute to him the cosmological quality of Tanen, whose main characteristic is emerging out of the water. Tanen's function of supporting the solar creator is also hinted at: he emerges towards the High Seat.

2: The Protector *who raises the spear, overthrowing the enemy (tw3 mcb3 ḥr sẖr ḫftjw).* In agreement with these appellations he is also called the *sgmḥ*-spear. As noted above, the protective aspect embodied in these weapons originates in Tanen. Further, the Protector carries the *s3-t3*-snake as a guardian on his head *(s3-t3 m s3w tp.f).* Finally, he is called *the image of He-of-the-radiant-face (snn j3ḫw ḥr)* and the *Ḥtr-ḥr*-falcon – a characterization which echoes the presentation of the Protector in the cosmogony text.

3: He is the god who opens the water *(ptḥ nt)* – another function behind which Tanen can be discerned.

With this series of epithets, the God-of-the-temple is identified with the productive earth – the irrigated land sustaining life.

The life-sustaining quality of the God-of-the-temple is repeated in other Edfu texts[31], and it can similarly be specified according to a fertility aspect and a martial aspect, as in one presentation of the god where he is said to have granted marshes, and supplies of bread, poultry and fish, as well as having smitten the Asiatics and felled the *Jwntj*[32].

The God-of-the-temple is – not unexpectedly – called son of Tanen[33]; his theophany is the temple conceived of as growing from the earth and as representing its produce: the manifested life-producing qualities of the father. He can be seen as a cultic version of Tanen – when his legend says that he *emerges out of the water towards st wrt,* it has in this case a reference to the temple with the High Seat. The God-of-the-temple is the life-giving and life-protecting god of the inundated earth in the form of the temple. In the theophany of the temple he incorporates the creative and sustaining powers of the earthgod. That this "son of Tanen" in his protective capacity is thought of as identical with Horus cannot be doubted.

This survey shows that the text presents a number of gods, a large proportion of which are clearly Horus-hypostases, while Re and Tanen figure independently for the most part, even if they too are ultimately identified with Horus as god of Behdet. The following characterization can be made as regards this

31 E. g. E IV 103; Mohiy el-Din Ibrahim 1979.
32 E IV 258,11–259,9.
33 E IV 259,5 f.

relationship between the god of Behdet and the many divinities that appear. A polytheistic concept of god is the prominent one in the cosmogony. The myth employs creators of diverse orders and protectors of diverse orders. Nevertheless, a henotheistic concept of god can be discerned underneath, even if it is not pronounced, since it lies in the nature of myth to personify and hypostasize and to conceive of the divine development as the development of a pantheon. But in its laudatorial passages the creation text sporadically approaches a henotheistic concept of god in accordance with which the creation story is a story about the god of Behdet. Also, in the legends of the figures in the relief to the text, a henotheistic tendency can be observed. In addition, indications of various kinds are scattered throughout the text that Horus is in the theophanies. As the *Ḥtr-ḥr*-falcon he protects the floater and is supported by it as well. In this manifestation, he is associated with the pre-stage of cosmos. Then, he is identified with the rising sun, the dynamic mover who accomplishes the transition from the dark chaos to the illumined cosmos. Further, when the temple is built for Horus as *Ḥr-3ḫtj,* he is again conceived of as a solar god. His connexions with Tanen are drawn through his protective functions as well as through his position as lord of the sacred place: as god of the temple in *Wṯst-Ḥr,* he appears as the God-of-the-temple.

Moreover, there is the tacitly presumed conventional knowledge of Edfu that the god of Edfu is the all-encompassing god, as the following apostrophe succinctly presents him, addressing him as *Lord of the sky, the earth, the underworld, the water, and the horizon (nb pt t3 d3t mw 3ḫtj)*[34]. The qualities that are in the cosmogony apportioned to the falcon-gods of the sky, the gods of the earth, the gods of the water, and the gods of the sun – in the forms of *Ḥtr-ḥr, Sj3* and *Bjk, Ṯnn* and *Jr-t3, Nwn* and *Ḥcpj, Cpj, Rc* and *Ḥr-3ḫtj* – are here all incorporated in the god of Edfu.

b) *The mythical theophany implies a theogony*

The comings of the gods entail concurring comings of cosmic phenomena, the implication of which is that the gods acquire cosmic being. They leave their state of hidden existence in chaos and achieve manifest existence in the cosmic phenomena. The cosmos is the manifestation of their having been created. The pre-elements of cosmos, the inundation and the darkness are, in the theogenetic perspective, the ultimate origin of the gods. Our text especially focuses on the inundation as *fons et origo.* Tanen, the divine

34 E VIII 65,4.

"father" *comes out of the water* and thus his "father" is the water. The theogenetic role of the inundation is elaborated by other Edfu texts, as in the following example, where it is personified and identified with Hapj – the life-producing and life-sustaining Nile, and in this form is seen as the origin of gods:

Hail to you, O Nun! (jnḏ-ḥr.k Nwn)
Hail to you, O Flood! (jnḏ-ḥr.k Nwj)
Hail to you, O Hapj, father of gods! (jnḏ-ḥr.k Ḥcpj jt nṯrw)[35]

1. The theogonic appearance

The phraseology employed by the cosmogony text to introduce the two-fold creation of gods and cosmos stresses the aspect of theophany: the gods are *seen* – they are not hidden anymore in the water and in the darkness – but have acquired visual form. Words like *being seen, being perceived,* and *look!* frequently mark the coming into being of the gods and the cosmic phenomena.

Thus, the reed is *seen* (181,13). The *Ḥtr-ḥr*-falcon is *perceived* above it (181,13). The floater of Horus is *seen* (182,1). *Look!* introduces the coming out of the Divine Ruler from *d3t* (182,5). The constellation of the Adze is *seen* (182,7). The name of the lord is *seen* (182,12). This is the name of the god of Edfu as solar god: Horus-Re – in whose light that which is on The Two Lands manifests itself, displaying the nature of its creator. The utterance of his name is equated with a theogonic *appearance:* the god who reveals himself in the uniting of the sun with the heaven of The Two Lands is invoked. When his name is heard, The-sun-has-united and That-which-is-on-The-Two-Lands *appear into being.* The theophany of this creator is the sun shining in heaven as well as everything appearing underneath.

Further, *Mss* is *seen* (182,15). *Look!* introduces the coming of the Protector, *Nṯr-ḥn,* when he appears as the image of the Shining-of-face, as the *Ḥtr-ḥr*-falcon (182,16). *Look!* precedes the opening of the canal-water (183,1). The coming out of the earthgod from the water is *perceived* (183,3). The strong arm, *ḫpš,* is *perceived* as the body of *Wṯt,* the Begetter (183,4). *Look!* precedes the coming of the image of the protector, *snn nhp* (183,5), which is the weapon of the Protector with the head of *Ḥtr-ḥr* on it. The *Sj3*-falcon *looked at Ḥr-sḫm-ḥr* (183,7). The *pcj*-land that can be tread on is *seen* (183,11). *Look!* precedes the appearance of the Place-of-*W3j* as the Place-of-the-throne (184,3).

35 E II 143,5–7.

Look! precedes the appearance of the rejoicing *Sj3*-falcon-of-the-place-of-building (184,4). The founders are *seen* in front of the builders (184,11). Horus is *seen* on the *pcj*-land which is protected (184,18).

From the great number of these expressions, we might safely infer that appearance is a constitutive characteristic of the creation: that which is seen has *come into being*; a god who appears is a god who appears into life – having made the transition from latent existence into manifest existence[36]. The kind of creation implied by the theophanies is thus the *ḫpr*-creation. As we have pointed out in an earlier chapter[37], the *ḫpr*-creation is not a creation from nothing but a transition from latent to manifested being. Within the chaos-cosmos frame of reference it refers to the coming out of that which lies hidden in the darkness of the night and in the water of the inundation. The act of *coming out* actually qualifies the *ḫpr*-creation. Also, *ḫpr* always denotes the creation of a concrete phenomenon.

The introductory apostrophe to the god of Behdet focuses on this theogonic implication of the theophany, when it states that: *You are the divine god who came into being at the first occasion (twt nṯr nṯrj ḫpr m sp tpj;* 181,10). The epithet of *divine god (nṯr nṯrj)* indicates autogenesis[38]. The words directed to the god of Behdet qualify him as the god who comes into being. The ensuing cosmogony can be seen as an unfolding of the god of Behdet. As more places come into being, the god expands and differentiates into a plurality of forms: the many gods which appear are his hypostazised qualities related to each other in a mythological conception of the diversified all-encompassing god. Also, some of his hypostases are said to have come into being through a *ḫpr*-creation: When Horus comes out from the darkness he acquires being in the Winged Disk, and the transition is designed by the word *ḫpr: When Horus took to flying the Flying Disk came into being: cpj n Ḥr ḫpr Cpj.* Similarly, He-who-opens-the-water, The-Lord-of-fear, *(Ptḥ-nt, Nb-nrw)* is said to come into being, *ḫpr.* – And the coming into being of the following protecting deities is denoted by *ḫpr:*

36 Alliot has a more abstract theory of the cosmogonical implications of *m ḫf* and *m sj3,* according to which the first step in the creation process is that the creator sees in his mind what he will create through formulating it in words which he utters – by which act the content of the word is realized (1966 p. 134 notes d and f). A creative procedure of this kind is documented in Egyptian religion (on the so-called Shabaka stela). Yet, because of the temple-cultic stress on *appearance* from the state of being *hidden* (see also p. 104 ff.), I choose to understand the stress on words denoting being seen on the background of the chaos-cosmos imagery employed here: what lies hidden in the water and in the darkness manifests itself and is seen.

37 See p. 68 f.

38 Assmann 1979 p. 24 n. 65.

The Male-arm came into being: ḫpr Ṯ3j-c,
He-who-⸢ḥnk's⸣-the reed came into being: ḫpr Ḥnk-twr,
He-who-is-in-front-of-the-male-arm came into being: ḫpr Ḫntj-ṯ3j
He-who-is-powerful-of-arm came into being: ḫpr Ḫpš-c.

The creator-gods and the protector-gods as well can be understood within this henotheistic-theogonical frame of reference as the cosmos-producing and cosmos-sustaining activities of the god of Edfu, multiplying as the cosmogonical process develops. Thus the creation is not solely the coming into being of Edfu, but also the coming into being of the god of Edfu. He constitutes this cosmos. The place manifests his diversified existence; the entire cosmogony is a dynamic display of the evolving divine presence.

The briefest and most Egyptian formulation of this relationship between creator and work of creation is found in another Egyptian document, namely Papyrus Bremner-Rhind. The formulation is a wordplay on *ḫpr, ḫprw (form of existence), and Ḫpr (Khopri:* the sungod in his form of the beetle); the wordplay conveys an imaginative exposition of the essential reciprocity between the coming into being of the creator and the coming into being of his work of creation. The words are put in the mouth of Khopri *(Ḫpr)* who is the creator in one of his solar forms, the form he takes when he comes into being every morning: *Recite: Thus spake the Lord of All: When I came into being, Being came into being. I came into being in the form of Khopri who came into being on the First Occasion ... I came into being in the form of Khopri when I came into being, and that is how Being came into being ...*[39].

2. The theogonic invocation

As the cosmogony was effected through the uttering of names so, too, is the theogony effected through the uttering of names. The instances are few – since the creation is primarily seen *qua* cosmogony and attention thus focuses on the coming into being of the cosmological places – but significant.

The names of the solar creator are uttered. When the Winged Disk comes as the Hovering-one to do homage to the Lord of *Ḏb3, the words passed and the Ba of the Place-of-spreading (-the-wings) heard the name of the Winged Disk.* The god of Behdet is magnified accordly (182,9–10).

The solar god appears as the lord of the sky when the name of Re is uttered: *"Who is our lord?" The Wing came who uttered the word. And the name passed, being seen ... The name is Re, Horus Re* (182,11–12).

39 Translated by Faulkner 1938 p. 41 ff.

Also, several of the protector-gods are explicitly said to appear by invocation: the image of the Shining-of-face as the *Ḥtr-ḥr*-falcon, the Male-arm, He-who-*ḥnk's*-the-reed, He-who-is-in-front-of-the-male-one, and He-who-is-powerful-of-arm are all called into being (184,7–10).

The mode of creation is, then, a transition into visibility, and invocations are instrumental in it[40].

c) *Conclusion*

According to the creation myth the world of Edfu is the place in which gods come into being, or – rather – is the evidence that gods have come into being: the world of Edfu is divine life that has manifested itself.

This essential connexion between the gods and the place implies a particular definition of divine life: the life of the gods is cosmic life. In this view the theology of Edfu is in agreement with traditional Egyptian theology. The gods of Egypt are not independent of the cosmos[41], they constitute it. *God* is no spiritual being abstracted from matter, living transcendentally in his own spiritual world away from the material world of man, though frequently visiting it – by some sort of incarnation. The coming of the ba of Horus is no *visit to* the world of man: it is the coming of the creative sun and it brings with it the world of man.

Thus the evidence given by this cosmogony seems to point to a certain conclusion as regards the ontological status of the Egyptian gods. The Egyptian gods differ from those found in religions with a dualistic ontology according to which god is spirit and the cosmos is matter, and spirit and matter belong to separate worlds, even though they can mingle. The Egyptian gods are not spiritual in this exclusive sense; they exist as cosmological phenomena or they do not exist at all. The same is valid for the Egyptian creator. He exists as cosmos or he does not exist at all. And he comes to life not from a world of his own, but from potential cosmos.

40 The feature is not exclusive of Edfu, see Morenz 1960 p. 24; Assmann 1969 p. 210 f.

41 See Frankfort 1948; Hornung 1973.

II. The temple-cultic theogony

a) The temple as a stage for the theophany

The coming-into-being of the gods is ritually staged as a cosmological theophany in which the cosmos and the gods are simultaneously revealed. The central divine figure in this cultic tableau is the solar god, acting as creator; and the scenery is the chaos-cosmos-imagery of the temple. The interior represents cosmos hidden in chaos: a place which has not yet appeared. The *intrada* of the god of daylight will change the hidden place into a place of cosmos *qua* pantheon: the cosmos will appear in the light together with the gods constituting it. In this ritually enacted *ḫpr*-transformation the dynamic character of the temple can be seen: it is no static image of the hidden cosmos, no petrified monument commemorating the "first time", but a manifestation of divine appearance.

Thus, as was the case with the mythical theogony, so is also the ritual theogony an *appearance* into being. The gods leave their state of latent being in chaos to be seen as cosmic phenomena; the mythical, theogonic theophany is paralleled by a ritual one – recurrently enacted, which shows that the creation of gods is not an affair of the past and bygone.

The appearance of the creator and the gods

When the god of Edfu comes as the solar creator, the gods of the cosmos of Edfu appear in his light. The time of creation is the dawn. When the morning light pervades the dark building, this place of chaos is changed into the place of the cosmos and simultaneously the gods belonging to the cosmos stand revealed – and in two modes – corresponding to the two metaphorical meanings of the temple: as dwelling and as the cosmos. The statues in the chapels are seen – the dwellers in the mansion; also, the reliefs of the gods are seen engraved all over the building, on its walls, in its ceiling, on its columns – the pantheon of the cosmos, its appearance coinciding with the appearing of the cosmos in this enactment of the theophany of the creator.

The ritual use of light

There are two ritual ways of bringing light into the temple: one is by opening its doors; the other is by artificial illumination.

The opening of the doors is a ritual event initiating the epiphany of the creator. According to the textual evidence all doors were opened at dawn[42] to let in the god of light that he might "unite with" his seat. The opening was accompanied by an act of adoration[43], which is one of the many indications that the letting in of the light manifests a theophany. And as in the cosmogony myth the praisegiving, *ḥsj-Rc,* was followed by the appearance of the cosmic places so is this divine coming and the adoration of the god followed by the cosmological transformation of the temple: its interior takes form. The *seat* with which he is said to *unite* can be defined as the High Seat, *St wrt,* a locality in the temple which occupies a central position in the theophany scene. The temple is constructed in such a way as to lead the light in a straight line into the *St-wrt*-sanctuary which lies axially at the beginning of the processional road through the temple from its innermost localities (plate 8). When the doors of the temple and the sanctuary are opened the light travels along this road and hits the solar boat resting on the representation of the mythical first mound emerging from the water and uplifting the god who creates the world with his rays. The ingoing light has a double function in that it both represents the uniting of the god with his seat and also effects the appearance of the creator from the place of darkness – in a reflection outwards along the same road from the sanctuary of the "beginning", the two directions (one being of the cosmic theophany, the other of the cultic theophany) converging along the road.

Thus the first mound of cosmos has appeared together with its god, the solar creator alighting on it after having been hidden during the night in *d3t.*

As regards the illumination of the places which the light from the doors cannot reach, the light is directed to these through an ingenious system of holes in the roof of the temple, which leads the light to the reliefs of gods thus made to appear, in a kind of "spotlight" illumination whereby the reliefs of significance for the cultic scene are discerned[44].

Artificial illumination is required in the chapels which are situated at the farthest end of the building behind the *St-wrt*-sanctuary, and which have no holes in their roofs. These places lie in darkness even when their doors are opened, for the sanctuary blocks the passage of the rays. Thus the chapel of *Msn* has to be illuminated artificially.

42 Fairmann has discussed this practice in "Worship and Festivals" p. 176 f.

43 Fairman, *op. cit.* p. 176–179. In hymns to the sungod, the appearance of the god in the horizon can be expressed through the image of opening gates or doors, see Assmann, e. g. 1970 pp. 21, 48.

44 Sauneron and Stierlin 1978 p. 88 ff.

b) *The revelation of the statue: the god living in his dwelling*

1. The daily appearance

The chapel of *Msn* contains Horus-statues and is an important place in the morning ritual. The illumination of the chapel is done with candles[45]. From the textual evidence, the lighting of the candle is a ritual correlate to the sunrise – the light of the candle represents the light of the sun: *It shines like Re rising in the horizon*[46], to quote a Karnak text which expresses a universal idea also found in Edfu.

In the statue-ritual performed in the chapel, the dwelling-imagery of the temple is apparent: the revelation of the statue is the beginning of the morning toilet which the god undergoes after having slept during the night[47], and it ends with his having breakfast.

But it appears from the texts pertaining to the ritual that the *face (ḥr)* which comes to light is not only that of the statue but the very face of the sun – the cosmic theophany of the creator. The Revelation-of-the-face-of-the-god *(wn-ḥr ḥr nṯr)* is a cultic epiphany parallelling the cosmic theophany[48]. Its performance is documented in the *St-wrt*-sanctuary as well, which also contains a statue of Horus – as the solar god[49]. Some examples can be quoted to illustrate the equation. From a text on the *St-wrt*-sanctuary the following words can be quoted:
Revealing the face, adoring the face. To be uttered: You arise as Khopri as you come out of Nnt [a]*, and your rays spread over the world! (wn ḥr dw3 ḥr ḏd mdw wbn.k m Ḫpr mj pr.k m Nnt wpš m3wt.k ndbt)*[50].

a) The night sky, pendant to *Nwn,* the chaos waters.

Similar words are uttered when the face is revealed in the *Msn*-chapel: *Shine on the earth as in the sky (wbn ḥr t3 mj ḥrt)*[51].

These words presuppose the *cosmos*-imagery of the temple; the revelation of the statue is given a cosmological meaning which goes beyond the dwelling imagery of the temple. Thus, in the ritual revelation of the face the two metaphorical aspects of the temple concur.

45 Alliot 1949 p. 62.
46 Alliot 1949 p. 63, ref. to Moret 1902 p. 9.
47 E I 35,4: The temple is "his great dwelling upon earth wherein he sleeps until dawn", translated by Fairman 1941 p. 426 n. 127.
48 Assmann 1969 p. 152 f.
49 E VII 15,3–7.
50 E I 40,16–17; pl. XII.
51 E II 87,13.

The aspect of revelation is stressed by the officiating priest when he approaches the sanctuary and presents himself to the god: *I am a prophet, the son of a prophet; it is the king who has commanded me to see the god*[52].

This theophany is enacted every day. In the morning the priest walks from the pronaos – where he has been purified – to the chapel, opens its doors and raises his candle. The candle contains perfume emitted while it burns, signalizing the presence of the god[53]. When the dark room is illumined by the candlelight, this means that the god of Edfu has appeared – both as creator and as created: he is the revealing light, and he is the god of the statue upon which his light falls: he is himself the creator of his Horus-form.

The statues of Horus in the chapels of *St-wrt* and *Msn* represent forms of the god known from the cosmogony myth. The statue of *St-wrt* represents him as the solar god of the "beginning". In the *Msn*-chapel there are two Horus-statues:

1: There is the statue of He-of-the-mottled-plumage *(s3b šwt)* in the form of the squatting *gmḥsw*-falcon, who is identified with the *Sj3*-falcon and also with Horus-of-the-horizon *(Ḥr-3ḫtj)*[54]. The statue is phallic and the end of the male member is shaped like a falcon-head[55], which makes it likely that there is a connexion between this god and the mythical He-who-is-in-front-of-the-male-one, while at the same time indicating that the protection performed by the latter has a life-sustaining meaning. In a Denderah-text he is named *Ṯ3j,* the Male-one[56]. The statue is paired with a statue of Hathor as Maat; she is considered his consort[57]. Together they represent the upholder of cosmos and the order of cosmos *(m3ct).*

2: The second Horus-statue is the Falcon-of-gold (*Bjk-n-nbjt)*[58]. The name attributes to the god a solar aspect – gold signifying the sun[59].

Thus the central Horus-hypostases of the cosmogony appear during the morning statue-ritual.

The weapons of the god also appear, and in two versions. In the *Msn*-chapel two cultic spears are apparently set up. Beside Harakhte (the *Sj3*-falcon, He-of-the-mottled-plumage) is *the staff of Horus of Behdet (p3 mdw n Ḥr*

52 E III 83,10.
53 Alliot 1949 p. 62 f. Cf. E 182,3, and note 29 in the commentary p. 30.
54 E IV 5,1–3; 13,7–9; VII 13,1–2.
55 E I 554,4–6; Alliot 1949 pp. 316, 318.
56 Chassinat, *Dendara* IV (1935) p. 20,3–5; cf. E VI 184,7–8.
57 E IV 5,1–3; 13,7–11; V 8,9; 10,4–5; VII 13,1–2.
58 E IV 13,10–11; Alliot 1949 p. 322.
59 Derchain 1962 c p. 71; Daumas 1956.

Bḥdtj)[60]. Beside the Falcon-of-gold is *the splendid sgmḥ-spear which comes out of the water (p3 sgmḥ špsj pr m nwn)*[61]. According to the reliefs in the *Msn*-chapel the spears are similar to each other[62]. Both have a falcon-head near the tip, like the weapon depicted in the relief to the cosmogony text on the enclosure wall. But the spear of Harakhte has in addition a solar disk on the falcon-head. A little piece of cord attached to the spears indicates that they are harpoons. The spears depicted are to all appearances actual cultic spears, as the reliefs present a cultic scene. On the basis of this text- and relief-documentation it seems as if we can safely conclude that there were two spears erected in the *Msn*-chapel, similar in appearance but differentiated according to the different forms of the god of Behdet.

There seems to be a connextion between this particular locality in the temple and a situation described in the cosmogony text:

> ⌜*When that which is in the beyond settled down*⌝*, (namely) the Sj3-falcon together with the image of the One-great-of-strength, the Sj3-falcon looked at Horus, powerful-of-face, (which is) the name of the* ⌜*weapon*⌝*. It is elevated, the sgmḥ-spear,* ⌜*as*⌝ *the Horus-of-the-sḫm-ḥr-weapon separated (from the water) by Tanen ... As the grand ḥḏ < of > the Maker-of-the-earth was brought to his son, Tanen brought and established the one which is similar before Horus-of-rejoice. (Thus) the strong arm is before the God-of-the-temple, the protection of the šbtjw* (183,7–10).

The words are strongly evocative of the *Msn*-chapel and its cultic equipment.

Horus-of-rejoice, *Ḥr-3wt-jb,* has its cultic counterpart in the Horus-statue having received the *3wt-jb*-amulette[63].

On the whole, there is an unmistakable cultic reflection in the theophanic imagery of the cosmogony. The cult-images of Horus make their impact felt in several of the descriptions of the god.

Perhaps the clearest example is found in the theophany connected with the break of dawn. When Horus appears from the dark underworld he is described as *the one with a head beautiful of face, the Hovering-one spreading real marvels, making the darkness m3c with his wings, (being) mailclad and (being) majestic, clothed in s3wj-gold* (182,4–5). There can be little doubt

60 E I 232,8.

61 E I 239,8.

62 E XI ph. CCIV; ph. CCXCVI.

63 Alliot 1949 p. 168 ff.; cf. also relief in the *St-wrt*-sanctuary E XI pl. CCXVII; E I 32,11 f., and representations in the treasury, E II 285,15 pl. XLVb.

as to the influence exerted by the cultic representations of the god in this description[64].

2. The theogonic nature of the daily appearance

According to the morning rites and the hymnic texts, the appearance of the god is interpreted as a creation or as a birth.

The god who appears is a god whose latent life in chaos takes form; he undergoes a *ḫpr*-transformation. This idea is expressed by the officiating priest: when he sees the statue of the Falcon-of-gold, he identifies the god as Khopri, *Ḫpr,* the creator who has transformed himself into being (in Alliot's translation)[65]:

Vision du dieu. Dire: Je m'approche de "(Celui qui est) intact", (pour) voir l'image glorieuse de Dieu, de mes deux yeux, (pour) contempler la statue de l'Ètre divin, la forme sainte du "Faucon d'Or!"

(m33 nṯr ḏd mdw wḏ3.n.j Wḏ3 m33.j sḫm ḥc nṯr m jrtj.j dg.n.j smn n Ḫpr nṯrj tjt ḏsrt n Bjk-n-nbjt)[66].

The cosmological form of Khopri is the golden, rising sun, as we have already seen in a hymn inscribed in the *St-wrt*-sanctuary: *You arise as Khopri as you come out of Nnt, and your rays spread over the world*[67].

Or, the ritual theophany is conceived of as a *birth.* Thus a hymn inscribed in the Hall-of-the-ennead outside the *St-wrt*-sanctuary greets the god in this way:

Praise to you, god of Behdet, lord of the sky,
splendid Cpj who shines in the Horizon,
you beautiful sun who illuminates the darkness,
splendid child who illuminates the earth,
pupil of the Wḏ3t-eye who illuminates The Two Lands with his rays.
... old man at night,
young child in the morning (ḫ rnpj m dw3w)[68].

64 As regards the theological and cosmological meanings of the minerals used in the cult images, see Aufrère 1982–1983.

65 Alliot 1949 p. 79.

66 E I 26,4–6; pl. XI.

67 E I 40,16–17.

68 E I 379,5–18.

This conception of the morning sun belongs to the mythological image of the night sky personified as a mother, cosmologically located below the earth, into whom the sun sinks at evening as an aged man dying[69] and from whom he is reborn in the morning[70], as an infant child – the god of light born of darkness. The theogenesis is thus placed within a time cycle of 24 hours.

The mythologem of the night sky as the mother of the day is one of the most influential in Egyptian sacred art and architecture. In the temple the topos of the night mother is in the ceiling[71] – the interior of the building representing the dark netherworld – and its mythical meaning ascribes to the dark room the meaning of the *fons et origo* of life.

Also, in the hymn one finds an obvious topographical double-reference to the horizon and to the sanctuary. The *horizon*[72] is located in the temple[73] at the end of the hymn, and the temple itself is named according to its metaphorical meaning of *Egypt*:

> O, you king, Horus of Behdet, great god, lord of the sky, you of the mottled plumage, *who comes out of the horizon in the Two-temples-of-the-South-and-the-North (... pr m 3ḫt ḫnt jtrtj Šm3w Mḥw).*

The sanctuary represents the place of the cosmological transition to life[74].

Thus, when the dwelling imagery of the temple is emphasized, the god goes to sleep in the evening and is awakened in the morning. When the chaos-cosmos-imagery is emphasized, the god dies in the evening and is reborn in the morning. The temple is the place of transition from chaos to cosmos, and correspondingly from death to life – the two ontological states are bridged in this place.

Another ritual expression of the theogonic aspect of the theophany is the embracing of the statue performed by the priest who, by this act, transfers

69 See Hornung 1972 p. 47 ff.

70 Cf. E III 50 (in Kurth's translation): *Tritt ein in den Mund als Sonnenscheibe mit einem Flügel. Deine Majestät möge wieder zwischen den Schenkeln hervorkommen (cḳ.k m r3 m jtn ḫrj ḏnḥ pr ḥm.k jmjwt mntj)* (1975 p. 41); and: *Nephtys, Worte zu sprechen: "Erscheine doch, komme hervor aus dem Schoss, Du, (der Du der bist) 'Der die Dunkelheit vertreibt, wenn er sich im Ostland zeigt'" (Nbt-Ḥt ḏd mdw ḫc rk pr m jḥtj ḫsr kkw dj.f sw m b3ẖ)* (1975 p. 42). It is a widespread hymnic motif in Egyptian traditions, see Assmann 1970 p. 20, and 1969 p. 120 f.

71 See note 45 in the commentary to part A.

72 *3ḫt* denotes the place where the sun *arises* (and sets), Wb I 17; cf. Brunner 1970 p.31 n.16. Assmann translates *3ḫt: Lichtland,* 1969 e. g. p.121.

73 For comparative material from the New Kingdom, see Brunner 1970 p. 31 ff.

74 Assmann 1969 p. 104.

the *ka*-life to the god[75]. The mythical image behind this rite is that of the creator who creates the gods by embracing them and thereby transferring to them the capacity to live[76]. The priest plays the role of the theogon.

3. The seasonal appearance

About forty festivals were celebrated in Edfu during the year[77]. From the point-of-view of ritual creation, the festival of Opening-the-year *(wpt rnpt)*[78] is the most interesting. It is an annual enactment of the two-fold creation of cosmos and pantheon at the beginning of the year. Also, it is the traditional time for the dedication of a temple[79].

The New Year festival is a festival which ideally coincides with the inundation taking place in July. Thus, the cultic beginning of the year is timed to the beginning of the natural year[80] – which is also the mythological time of beginning according to the cosmogony texts; the New Year festival is a festival of the *first time.* It also coincides with the rising of the Sothis, called Opener-of-the-year *(wp rnpt)*, and this star came to be recognized as the herald of the inundation[81]. Rites are performed during the festival which reflect the significant role played by the inundation. Thus in the doorway of the Chamber of the Nile is inscribed a text which mentions a ritual filling of jars with inundation water (in C. Traunecker's translation: *Je t'apporte le vase nemeset vénérable (rempli) de ce qui sort du Noun, l'eau nouvelle qui se lève dans le lac (?) afin de te saluer pendant les jour epagomènes, quand Nout a donné naissance à ses enfants, ainsi que le jour du Nouvel An lorsque Re est sorti du Lotus dans le lac*[82]. This presentation of the new water takes place – according to the text – during the five intercalary days introducing the festival; they are called *those upon the year (ḥrjw rnpt)* and are added to the year[83].

75 Alliot 1949 p. 81, ref. to Moret 1902 pp. 79–102. Also the rite of fastening the broad necklace on the statue can be mythologically interpreted as the enlivening ka-gesture, E. Otto 1958 p. 13. For this meaning of the rite in mortuary context, see E. Otto 1960 scene 54.

76 Cf. Pyr. 1652; Frankfort 1969[6] p. 66 ff.

77 Fairman 1954 p. 182.

78 Wb I 305,1.2; Alliot 1949 pp. 303–433; Fairman 1954 p. 183 ff.

79 Blackman and Fairman 1946; Fairman 1954 p. 187.

80 The Egyptian concept of year is intimately connected with the annual inundation, see E. Otto 1966 p. 746 ff.

81 Parker 1950 p. 31 f.

82 E II 238,6–8; Traunecker 1972 p. 233.

83 Wb II 430.

Libations are the usual ritual representation of the creative Nile-water. As such it is also presented in the relief to the long cosmogony text; the relief can be understood as a ritual version of the cosmogony – which is an inundation story. It depicts a scene of adoring the falcon of Edfu by the God-of-the-temple behind whom stands the libating Hapi, identified with the Edfu-canal. This libating figure personifies the Nile precisely in its aspect of inundation[84].

The theogonic character of the New Year festival

An important feature of the text from the Chamber of the Nile quoted above is that the creative capacity of the Nile is seen in terms of a theogony: Nun is the father of the gods, and Nut his consort who bore them. Here is a well-known mythologem timed to the five intercalary days: during these days Osiris, Horus, Seth, Isis and Nephtys are born[85] – they represent pantheon conceived according to a sociomorphic model – that is the family – and the chaos elements of the inundation and the dark night sky are their parents.

There is also another divine birth connected with the festival, namely that of the creator; his birth precedes that of the pantheon. The last month of the year is called The-birth-of-Re *(Mswt Rc)*[86].

The theogonic role of Re is a central one in the festival. It is enacted in the form of a *ḫnm-jtn*-rite. The rite is performed for the benefit of all statues in the temple. It is enacted on the roof of the temple where the statues are brought in a procession[87] – starting in the dark temple and coming out into the light of the roof. The statues heading the procession are the phallic Horus-statue and the Hathor-statue of the *Msn*-chapel, the cosmos-sustaining and cosmos-controlling gods making their appearance. Exposed in the sunlight on the roof, possibly in a special chapel constructed for the purpose[88], the face of the statue is revealed, analogous to the daily revelation of the face, and the sungod *unites with (ḫnm)* it. The rite is called *ḫnm-jtn*[89]or, alternatively, *m33-jtn: Seeing-the-disk*[90], a designation alluding to the revelatory character

84 De Buck 1948.

85 Alliot 1949 pp. 237 ff., 261; Gutbub, *Kom Ombo* I (1973) pp. 21, 29; Sauneron, *Esna* V (1962) p. 28; Schott 1950 p. 992 f.

86 Wb II 141,13. *Ad* the discussion concerning the relationship between *wp rnpt* and *mswt-Rc,* see Parker 1950 p. 46 f.; Gardiner 1906 p. 142; Brugsch, ZÄS 8 (1870) p. 109.

87 E I pl. XXXVII.

88 Alliot 1949 p. 342 ff.; Fairman 1954 p. 184. See plate 7.

89 E I 549,7; Alliot 1949 pp. 308–428.

90 E I 441,9.

of the creation. The rite seems to have been timed to coincide with the sun in zenith[91].

The seasonal appearance of the creator

A processional emergence of the sungod from the temple is staged at certain festivals. The boat of the solar creator is carried on the shoulders of priests[92] (who are his cultic crew) from the *St-wrt*-sanctuary, the cultic correlate to the cosmological starting point of the cosmological movement of the sun-boat. It is carried through the halls of the temple as if sailing on the waters of the dark underworld, and comes out into the daylight of the open forecourt, as if sailing on the waters of the sky of the day. In this route leading through the temple there is a coalescence between the chaos-cosmos-imagery of the river Nile and another cosmological water-imagery: according to a traditional image of cosmos it is surrounded by waters – the water of chaos below and the water of cosmos above; on these waters the boat of the sungod sails, on a route which thus comprises the night sky of the region below, and the day sky of the region above. This cosmological route has its ritual counterpart in the route through the temple. The procession starts in the dark interior of the temple, and when moving inside the building the boat sails on the waters of chaos. It proceeds to the open forecourt where it is fully revealed on the cosmic waters. The water-imagery of the forecourt[93] is enhanced by the decorations on the base of its walls presenting reliefs of processional boats, as well as real sacral boats sailing on the river[94]. On some occasions the procession left the temple for destinations outside Edfu[95].

The processional appearance of the creator is in the cultic vocabulary denoted by the words *prj* and *ḫcj*. *Prj*[96], *come out,* is a word which also frequently occurs in the cosmogony text where it denotes the appearance of gods. It is complementary to *ck*, *enter,* used to denote the god's entry into the temple. *Ḫcj, appear,* is a usual designation of the procession. It is a central rite, which is reflected in the word for festival: *ḫc* – derived from *ḫcj*[97]. The importance of the *ḫcj*-procession is also attested by the temple names and

91 Fairman 1954 p. 186.
92 E II pls. XL f, XL g.
93 The design of the court is that of a pond surrounded by vegetation. Sauneron and Stierlin 1978 pp. 26, 37.
94 E XIII pls. CCCCLI, CCCCLII; E X pls. CXXI, CXXII, CXXVI, CXXVII.
95 Alliot 1944 p. 535 ff.; Fairman 1954 p. 197 ff.
96 Wb I 519,18.
97 Wb III 241,6; cf. III 239 f. See Morenz 1960 p. 94 f.

names of the pronaos and forecourt: the places where the god appears. In a hymnic address the temple is called *Pr-ḫcj*[98]: House-of-appearance. The pronaos and the forecourt can both be designated *wsḫt-ḫcjt:* hall-of-appearance, court-of-appearance[99].

The *ḫcj*-function of the temple is primarily connected with the coming of the solar god into the open, but it might not be unlikely that the procession inside the temple has its own significance: it is probably associated with the mythologem of the sungod coming to the gods of the underworld[100]. The gods carved on the walls and pillars appear in the light from the torches of the procession.

In the cosmos-sustaining perspective, the solar boat of Re possibly amalgamates with the mythic boat of Horus in which this protector-god sails down the river to slay the enemies. The function of his boat corresponds to a similar one ascribed to the solar boat: when the latter is in the underworld it is attacked by Apep, the snake representing the darkness of the underworld, and only when this enemy of the light is vanquished can the boat proceed and rise in the horizon[101]. Thus both boats carry the god of cosmos in his fight against chaos.

4. The latent life of the gods is located in the temple

Behind the rites enacting the *prj* and *ḫcj* of the creator lies the concept of *latent being.* In Egyptian terminology this state of latency is qualified as a state of being *hidden: jmn.* Through the rites of *prj* and *ḫcj* the creator leaves his hidden being in the dark underworld and appears in the manifest life of daylight. The designation *hidden* can be applied to the temple as well as to the god. When the temple is seen as the topos of the latent being it is called the *hidden (place)*[102]. In a text inscribed on the sanctuary of the High Seat the sanctuary is given the qualification: *the High Seat hidden (jmn) in Wṯst*[103]. When the god is in his state of latency he is designated *hidden,* as in the following address to the High Seat made by Re: *O you Seat ... I have made you my House-of-appearance in the Mansion-of-the-Falcon, as I have hidden*

98 E IV 2,2; 319,4.
99 Wb I 366; E III 355,6; V 5,5; Fairman 1954 p. 168; Arnold 1962 p. 94.
100 E VIII 91,15–92,2; 92,4–6. See also Hornung, *Das Amduat* I p. 18 ff.; Assmann 1969 p. 150.
101 Assmann 1969 pp. 83, 277.
102 E IV 7,6.
103 E I 13 right.

myself in you[104]. In this quotation *appearance (ḫcj)* and *hidden (jmn)* correspond to two functions of the temple and to two states of the god as well: the states of latent and manifested life. As a last example can be noted the following excerpt from a text on the enclosure wall, in which both god and place are designated *hidden: the gods are in their chapels, the ennead is in its hall, the Hidden-one is hidden in the Seat-of-hiding* (nṯrw r sḥw.sn psḏt r wsḫt.sn Jmn jmn m St-jmn)[105]. The place of the Seat-of-hiding is in this instance probably the *St-wrt*-sanctuary – fronted by the Hall-of-the-ennead and surrounded by the chapels of the gods.

There is also another word which is often used to denote what is latently existent, namely *št3*. The word is usually translated *secret, mysterious*[106], but this does not give the essential meaning. It can also be translated *hidden,* and I prefer this meaning in our context. The word refers to the state of latency in the sense of not yet being revealed[107]. The word can be applied to objects as well as to gods which are in the state of being hidden. The shrine containing the statue of the god can be termed *št3,* e. g. *his hidden and grand naos of black stone (k3r.f št3 wr m jnr km)*[108]; the naos is hidden in the darkness of the chapel. Similarly the statue is hidden: *his hidden image in the Msn-chapel (bs.f št3 ḫnt Msn)*[109].

Jmn and *št3* thus refer to the *d3t*-aspect of the temple and what is in it: the temple represents the place of latent life. To obtain a better understanding of the ontological qualities of the underworld we should consult the literature which has been especially written about it, e. g. the *Amduat* or the *Book of the hidden room,* the *Book of the gates,* the *Book of the night* and the *Book of the day.* This literature gives the other half of the story of the cosmogony, namely a description of the domain where the gods of cosmos are latently existent and from which they appear: the place of the beyond.

The main theme of this literature is the sungod's passage through the underworld – from the place where he enters in the evening to the place where he comes out again in the morning as the Winged Disk identified with Khopri[110], praised by gods who sing his hymn[111] – the mythical correlate to the cultic hymn-singing in the temple at the same moment.

104 E VI 319,4–5.
105 E VII 12,4 f.
106 Wb IV 551.
107 Wb IV 551,14.
108 E IV 15,1.
109 E I 371,3–4.
110 Hornung, *Das Amduat* I p. 192.
111 *Op. cit.* p. 195 (834–845). Cf. also *ibid.* 1979 b, and 1981.

The description of the underworld offered by this literature characterizes it as the place of latent life in the same terminology as that used of the temple. Thus it is a world which is *hidden (jmn)* in the darkness. It contains all components of the world that has *appeared:* waterways (the god passes through this realm in his boat), marshes and fields; but here this Egyptian landscape is *not seen.* The *ba*s of the underworld are likewise *hidden (št3)* and so is the creator himself: his form is not seen. Sometimes his lack of form is conceived as a snake, the traditional "form" of what is formless in Egyptian iconology and mythology[112]. Even the creator's boat takes on the snake-form[113]. The dark underworld is the realm of snakes. There is also the snake which lives from the speeches of the gods[114], and there is the snake which is named "Life of the gods"[115], and there is the snake which swallows the image of Atum[116] and lives from the shadows of the dead and of their corpses. This food of the snake indicates the identity of the latter: It "is" the hidden gods. Thus, the texts state that the gods will again emerge from the snake[117], and so will the boat of the creator[118].

The appearance of the gods from the state of being hidden is effected when the gates of the underworld are opened[119] and the hidden room is enlightened – in a manner evocative of the appearance of the gods in the temple, when the gates are opened to let in the light.

Another feature of significance from our point of view is the emphasis placed on *knowing the names* of the hidden places and the hidden gods belonging to them. This knowledge bestows upon those who are in the underworld the power to *go in and out* of it, as the *Amduat* puts it[120], the underlying thought apparently being that he who recites the names of these places "calls" them into being. Here there is a parallel to the notion of the creative words of the Great Flood. Also, the cosmogony myth presents the appearance of the gods as effected through name-recitals, on the analogy of the *Amduat* which says that the creator calls out the names of the hidden *ba*s in the underworld, and they follow him to the horizon[121] – that is the place of transition from the

112 Hornung 1956 p. 31 f.
113 Hornung, *Das Amduat* I p. 68 f.
114 *Op. cit.* I p. 73 (323).
115 *Op. cit.* I p. 199 (869).
116 *Op. cit.* I p. 181 (756).
117 *Op. cit.* I p. 198 f. (857–868).
118 *Op. cit.* I p. 197 f. (856).
119 The Book of the gates, see Hornung 1972 p. 306.
120 *Das Amduat* III p. 35.
121 *Op. cit.* I p. 171 (710–717).

state of latency to that of appearance. He is the god whose voice is heard before he is seen, as the smaller cosmogony text on the enclosure wall characterizes him[122].

It follows that *jmn* and *št3* have precise ontological references. When they are epithets of the creator, they do not thereby allude to man's inability to grasp god's secret nature, the mysteriously hidden god whom man cannot know[123]; this would be an interpretation which is not really to the point, as it is too anthropocentric in its perspective for these mythical-ritual texts. The *hidden god* belongs basically to an ontological category.

This interpretation of the concept of the *hidden god* differs from that of J. Assmann who sees in the epitheths *jmn* and *št3* an indication of the "Übernatürlichkeit" of God[124]. Also, he finds witnessed in them "die Idee der *Verborgenheit* Gottes, sowohl im absoluten Sinne theologischer Unbestimmbarkeit, als auch im Sinne der Verborgenheit gegenüber der Götterwelt. Der Weltgott ist zugleich und als solcher ein *deus absconditus.* Die Einheit des weltgewordenen und daher vielheitlich manifesten Gottes ist ein undenkbar Geheimnis, an dem jeder geistige Zugriff scheitert und das auch den Göttern verborgen bleibt"[125]. It is, however, open to question whether Assmann has not here reasoned too much in the fashion of Western culture. The frame of reference for the interpretation of the terms *jmn* and *št3* is all-important. In the examples given by Assmann the frame of reference seems to be the latent being of the creator and not the deep mystery that cannot be grasped; the hymns quoted to support his interpretation present the god as the creator who comprises the latent being of the gods and of the world. Even the hymn of Pap. Leiden J 350 IV (17–19) should be understood in this way. It says (in Assmann's translation):

Einer ist Amun, der sich vor ihnen verborgen hat,
der sich vor den Göttern verhüllt, so dass man sein Wesen nicht kennt.
Er ist ferner als der Himmel, tiefer als die Unterwelt.
Kein Gott kennt seine wahre Gestalt.
Sein Bild wird nicht entfaltet in den Schriftrollen,
man lehrt nicht über ihn (in den Tempelschulen).

122 E VI 15,3; 17,6.

123 Unlike the view expounded by Assmann in "Primat und Transzendenz" p. 33 ff.

124 *Op. cit.* p. 26.

125 *Op. cit.* p. 33; cf. also p. 39, and his "Grundstrukturen der ägyptischen Gottesvorstellungen".

Er ist zu geheimnisvoll, um seine Hoheit zu enthüllen,
zu gross, um ihn zu erforschen,
zu stark, um ihn zu erkennen.
Man fällt tot auf der Stelle vor Entsetzen,
wenn man seinen geheimen Namen wissentlich oder unwissentlich ausspricht.
Es gibt keinen Gott, der ihn dabei anrufen könnte,
Ba-artiger, der seinen Namen verbirgt wie sein Geheimnis[126].

The god who is "vor ihnen" and who is "ferner als der Himmel" and "tiefer als die Unterwelt" is the god who comprises pre-being and the realm which is outside the manifested world: *the beyond,* the chaos. His chaos-side is not seen or heard, i. e. presented by the images or by the cultic words. Also, the hymn presents him as his own creator: no god invokes him – he comes as the first one and by his own mighty ba-power. Thus the text conceives this god rather as the powerful creator who is from before creation, than as *deus absconditus.* As Assmann points out, his name *Jmn-rn.f: He who hides his name,* is found already in the Pyramid-texts. These are texts that deal with the latent life of the underworld, and all through the literary history of Egypt the books on the underworld give the clearest explications of the meaning of the epithets *jmn* and *št3*. Thus, the following significant words ending the *Amduat* should be compared with the hymns to the creator (in E. Hornung's translation):

Der Anfang ist das Licht,
das Ende ist die Urfinsternis.
Der Lauf des Sonnengottes im Westen,
die geheimnisvollen Absichten, die dieser Gott in ihm verwirklicht.

Der erlesene Leitfaden, die geheimnisvolle Schrift der Dat,
die nicht gekannt wird von irgendeinem Menschen, ausser vom Erlesenen.
Gemacht ist dieses Bild dargestalt
im Verborgenen der Dat,
unsichtbar, nicht wahrnehmbar!

Wer diese geheimnisvollen Bilder kennt, ist ein wohlversorgter Verklärter.
Immer geht er aus und ein in der Dat,
immer spricht er zu den Lebenden.
Als wahr erprobt, Millionen Mal![127].

126 Assmann 1979 p. 33.
127 Hornung, *Das Amduat* III p. 35. Compare E. Otto 1960 scene 74 A.

It is against this background that the hymns to the creator as the hidden god should be understood. When Amon-Re ist praised as *Ba mit verborgenen Gesichtern und gewaltiger Hoheit, der seinen Namen verborgen hält und sein Bild geheim, dessen Gestalt man nicht erkannte am Urbeginn*[128], the reference is to the pre-cosmic creator who is latently existent from before creation. That this hidden god is called *ba* is to be expected, as *ba* connects god with his pre-being[129]. One can, therefore, agree with Assmann when he says that "Die Welt ist die sinnlich erfahrbare Manifestation und Verkörperung einer verborgenen Macht, die man als solche ba nennt"[130]. Also, he is right when he sees in these hymns a "Dialektik von Offenbarkeit und Verborgenheit"[131]; but these terms should be understood as referring to ontological categories.

The hidden creator has his cultic topos in the temple. When the temple closes its doors for the night, it becomes the *St-jmn,* the place where the god has hidden his form. The dark temple represents the underworld where he has changed his form into one that nobody *knows* and no god *sees,* as the *Book of the dead*[132] describes his latent being. *Know* stands here parallel to *see.* When the god is in the underworld, he is not known; when he has transformed himself into cosmological manifestation, he is known.

Thus the ritually enacted sequences of *coming out of (prj)* and *going into (cḳ)* the temple imply the theology of the creator who follows the dynamic movement of life – they correspond to the going *into* and *out of* the underworld. He is the *divine god (nṯr nṯrj)*[133] appearing from his latent being, but also, he is the god who vanishes into latent being again. In the daily and seasonal *coming out* and *going in* of the god is displayed the autogenetic principle so influential in Egyptian thought. About it E. Hornung has said: "Kaum eine andere Kultur hat das Nichtsein und seine schöpferische Potenz so vollkommen in ihr Leben integriert wie die ägyptische, hat so das Nichtsein bejaht ..."[134]. I would, however, like to modify his terminology. Instead of "Nichtsein" I would prefer "pre-being" or "latent being", because *being* does not emerge from *non-being* according to this ontology, creation is not creation *ex nihilo,* but a transition from that which is latently existing. In this monistic ontology, *being* originates from its own source: potential being.

128 Translated by Assmann 1979 p. 35.
129 See p. 135.
130 Assmann 1979 p. 35.
131 *Op. cit.* p. 34.
132 The Book of the Dead 175, cf. Hornung's translation in *Der Eine und die Vielen* p. 157.
133 Cf. the introduction to the long cosmogony text: E VI 181,10.
134 Hornung, *op. cit.* p. 179.

This autogenesis of life is impressively expressed in mortuary context. *Death* is life in the *št3*-state; and there is also a *coming out (prj)* from the tomb[135] analogous to the coming out from the temple; the function of the tomb is in this respect parallel to that of the temple: the tomb is a place of the hidden world from which the *ba* of the dead person comes out[136], the place of his latent life. Like the temple, the tomb is also a place of *uniting* with the sun: the sun unites with the reliefs or paintings of the world represented in the hall, when the tomb is opened – and the cosmic life of the dead person appears into being. He even has a statue which lives his *ka*-life in the cosmicized tomb.

The latent life of god, too, can be conceived of as his death. This is inherent in the image of the birth of the sunchild every morning: it presupposes the death of the aged god in the evening, and within this imagery the temple functions as his tomb[137]. There is also an explicit reference to the death of god in the form of a mortuary cult. Near Edfu there was a particular cult of the dead creators at the place where they lay buried[138]. There were similar cults of the dead creators all over Egypt[139]. The notion of dead gods should not be approached from modern Western conceptions of life and death according to which death is the final end of life. These mortuary cults are not for reflective commemoration of the most ancient gods, past and bygone, but for the communication with the gods whose life is going to manifest itself. The dead gods are the gods that will appear: they are the lords of latent life. Their qualification of being dead and buried is their trade-mark, so to speak: they are by definition the gods of the coming creation.

Thus the gods partake in the cycle of latent and manifested life. They are not outside it – they do not transcend life and death but have integrated both states into their being.

135 Faulkner 1972 Spells 1, 2, 9, 10, 12, 13, 17, 72; Hornung 1979 a and 1956.

136 The Book of the Dead 89, 92; see Kees 1956² p. 40.

137 Or, the relationship between the tomb and the temple might be understood the other way round: the tomb functions as a temple, i. e. the place where the transition from latent to manifested life takes place.

138 Reymond, MO pp. 117, 267 with references; Fairman 1960 pp. 87, 89.

139 According to Theban traditions, the ogdoad of Hermopolis went to Medinet Habu where they died and lie buried, Sethe 1929 § 102; Sauneron and Yoyotte 1959 p. 44. – North of Esna there is a mound where the seven *ḏ3jsw*-divinities lie buried, Sauneron 1958 p. 274. – The phenomenon is also documented for Philae, Junker 1913.

c) *The revelation of the temple-building*

1. The appearance of the cosmos-constituting gods

Not only the statues in the chapels are seen during the morning ritual. When the interior of the temple is illuminated by the opening of the doors and the lighting of candles, the entire place reveals what is hidden in it: the colourful world of gods: the illumination of the building brings into light the figures of the gods carved on it, i. e. it is a coming into being of the pantheon.

The admittance of the light has for its mythological correlate the marvellous theophany of the sungod – the Ruler-of-flying who comes out of the underworld as the ba of Horus, the Winged Disk in whose light cosmos stands revealed. It is a theophany which comprises chaos and cosmos; the Ruler-of-flying stretches out his wings before he is *seen*: he is hidden in the darkness of the night/the underworld; when he starts flying, he leaves his state of hiddenness and manifests himself as the Winged Disk. Also when staged as a ritual event, the theophany comprises both ontological states. The underworld from which the ba of the creator appears is, as we have seen, identified with the dark interior of the temple; the form of the winged sundisk is used decoratively in a way which is most suggestive of the mythological image, in an interplay with the dark room. Into the ceiling the skygoddess is cut, representing Upper and Lower Egypt, alternatingly with the Winged Disk in a succession of reliefs along the axial road through the temple from the *St-wrt*-sanctuary – the god stretching out his wings before being seen; he is with the goddess of the night sky[140]. His appearance is similarly staged in a combination of architecture and decoration: over the doors on the same axial road the image of the Winged Disk is engraved, in a succession of reliefs on these places of appearance. As the height of the doors is made gradually higher as the road approaches the forecourt – the cosmic room widening as the god flies – all reliefs can be seen when viewed from the forecourt when the doors are opened – with the effect that the Winged Disk seems literally to fly out from the underworld when the ingoing rays of the sun are reflected by the reliefs. They seem to be emerging from the dark interior[141]. In this way the ingoing cosmic theophany of the sungod is made to coincide with the outgoing

140 E II 130 f. pl. XL I: cf. the Morning hymn on the *St-wrt*-sanctuary: *Thy two wings of Him-of-Behdet, wherewith thou flyest through Naunet,* E I 16, translated by Fairman 1941: In this hymn Naunet has the double reference to the underworld and to the temple interior.

141 E III pl. LIII (pronaos); E III pls. XLVI–XLVII (doorway leading to 2. hypostyle hall); Sauneron and Stierlin 1978 p. 119.

cultic epiphany; by this ingenious decorative feature the notion is conveyed of the Winged Disk coming out of the underworld *united with* his image.

This is the appearance of the god in whose light the other gods appear too; the numerous gods of cosmos are engraved all over the temple – the pantheon identified with cosmos.

Even the outside walls of the temple are seen to be covered by gods in the sunlight[142]. Already the high enclosure wall proclaims the fact that it is a world of gods which appears. Huge divine figures immediately meet the eye. On the base, the Nile-gods of the nomes of Egypt walk in procession, carrying their produce as food-offerings. Rising above them, the main gods of the Egyptian pantheon cover the vast surface – gods standing and acting or sitting while receiving their offerings. The god of Edfu is represented in different shapes – as the warrior slaying his enemies, as the seated king-god receiving adorations, as the falcon hovering with outspread wings, or as the winged sundisk in the roof and over the lintels. The temple appears as a profuse compilation of gods. In principle, all gods of Egypt are carved into the building[143].

From the point of view of its reliefs, the temple building is thus the most imposing exposition of the intimate connexion between the cosmos and the gods. When the temple appears as cosmos it means that it appears *qua* pantheon.

Also in this temple-cultic perspective, the relationship between the creator and the concrete object is expressed as a *union*. The creator *unites with* the temple:

> *His ba having come, it united with his temple*
> *(jj.n b3.f ẖnm.f ḥt-nṯr.f)*[144].

The theogonic consequence of this *uniting with the temple* is staged as a tableau: the pantheon carved into the temple appears as united with its creator – which means, it has been created. The effect of the uniting of the ba of the god with his temple is that all gods are seen in his light and acquire existence in his theophany. The mythologem of the ba flying out of the underworld and spreading his light in the cosmic room is his coming to unite with the divine forms of cosmos, and these include not only the forms of the Winged Disk – engraved over the doorways along his road – but the forms of all gods: they come into being in his light and they are forms of his light, manifesting his diversified being. The uniting act of the creator means that he

142 E IX pls. IV, IX.
143 E III^2; X; XI; XII; XIII; XIV.
144 E IV 11,5.

communicates himself; the gods receive his ba-capacity, the capacity of the light manifested in the sundisk – which is the power to appear from the dark underworld.

Thus, even though *b3* is often translated *soul* it has to be emphasized that this is not an equivalent to the immaterial soul of Western religion[145], the ba of the creator is no invisible soul joining the visual, material bodies of the gods. The ba that comes is the visual, living sundisk. For – nor is the sun lifeless material. When it is called the *ba of Horus* it is conceived as something more than a mere material phenomenon. The designation denotes it as a monistic entity of qualities which in modern Western thought are termed soul and body, spirit and matter, qualities which are clearly distinguished even if they can unite. It is therefore difficult to find a relevant synonym for the *ba*-concept in the vocabularies of the Western world. We are here confronted with one of the barriers to an appropriate explanation of Egyptian ontology and theology.

We can, however, formally ascertain that when the ba of the creator *unites with* the temple he unites simultaneously with the pantheon – which is *hidden* – and the latter comes into being as the temple appears. The relationship between the god and the gods is 1) a relationship of creation: the gods assume manifest existence in his light, and 2) a relationship of partial identity: the gods partake in his light and can be seen as forms of his being – in accordance with the henotheistic connexion between the creator and the gods perceived in the cosmogony myth. As the temple evolves in the light of the creator, his divine presence acquires shapes that display his qualities. With the coming of the creator, the temple is transformed into the multifarious form of the creator.

2. The theogonic invocation

As has been demonstrated in an earlier chapter, the creative words in the cosmogony myth have a cultic-ritual character: they are words to be recited and they are recited within the context of praisegiving. This contextual position implies that they have a theogonic effect – they call into being the places in which the gods have their being. The places are praised as divine. In addition, there are explicit invocations of the creator in the myth: his name is uttered and it brings about the cosmological theophany of the god. Praising the god by uttering his name is an act of magnification in which the god appears

145 See note 38 to part C.

according to the content of the name[146]; this form of creative worship is one of the many features which links the mythical creation with the temple – the place which is especially created for the worship and magnification of the god. The mythical uttering of the name of the creator corresponds to the ritual adoration taking place at the moment of the theophany and in which the name of the god is uttered. Thus there is a *ḥsj-Rc* performed in the morning when the doors are opened. The god is greeted with a hymnic address, such as the following one inscribed inside the *St-wrt*-sanctuary – the sanctuary of the god of Edfu as the solar creator of the first mound: *Praising the god! Four times. Recite: I praise your Majesty with the choicest words, with the formulas which magnify your esteem, with the great names, your sacred forms* ⌈. . .⌉ *at the first occasion (dw3 nṯr sp fdw ḏd mdw dw3.n.j ḥm.k m ḏ3jsw stpw m s3ḫw n (s)wr šfjt.k m rnw.k wrw ḫprw.k ḏsrw sp.n.k r.sn m sp tpj)*[147]. Moreover, the creative function of the names which magnify the majesty of the god is here reflected in the reference to *sp tpj,* the mythical time of coming-into-being.

Recital of names is a basic form of praisegiving, and its creative motivation has since long been recognized. The hymnic enumeration of names is a calling into existence[148], an invocation to god to manifest himself according to the qualities indicated by the names. The ritual invocation effects the appearance of god[149], by calling forth his hidden being – it effects a *ḫpr*-creation[150].

The creation performed in the temple is one of transformation. Also on this point there is a correspondence between the creative procedure adopted by the temple ritual and that of the myth. In the myth, the creative words uttered transform the place of chaos into the mounds on the *pcj*-land of Edfu; and during the morning ritual the temple is transformed into the place of gods, in a hymnic recital in the course of which a metamorphosis of the building takes place: its divine forms appear from where they have been hidden, and they are seen to constitute the building.

In agreement with their god-evoking intention, the hymns of the temple are generally inscribed on the doorways through which the creator makes his epiphany[151].

Thus, as the geographical texts, the donation texts and the building texts make explicit the actual aspect of the creation with regard to the place of god, so the hymnic texts make explicit its actual aspect with regard to the god of

146 E. g. E VI 182,5–13.
147 E I 41,10–12; Alliot translates *sp.n.k r.sn: en lesquels tu t'es révelé,* 1949 p. 80.
148 Assmann 1975 a p. 26.
149 *Ad ḫcj* as a hymnic motif: see Assmann 1969 p. 261 f.
150 *Ad* transformation as theogonesis: see Assmann 1975 b p. 22.
151 Assmann 1969 p. 367.

the place. They witness that the coming-into-being of the creator and of the gods is an everyday occurrence; god appears every morning. Also, they witness the mutual relationship between the gods and the place. This relationship is, in a particularly insistent way, set forth in the morning hymn engraved on the façade of the *St-wrt*-sanctuary. The hymn expresses systematically the notion that the cosmos is constituted by the gods and also that this living and divine cosmos is the creator himself – the gods are seen as members of his body. The *St-wrt*-sanctuary is, more than any other place in the temple, the place of the creator: it is the place where the sungod first appears. It is natural that it is also a place of hymnic invocations and that on its façade a most emphatic call to the god of Edfu to arise[152] is inscribed. The hymn was presumably recited or sung when the doors of the sanctuary were opened. It is inscribed on the side which turns towards the Hall-of-the-ennead, and the hall "renferme ce chant" – to express it in Alliot's words: the hall contains three other hymns giving praise to the risen sungod[153].

The form of the hymn is a list of names and epithets by which the god of Edfu and his co-dwellers are called upon to awake. But it is more than that. It is also a hymnic explication of the cosmos as the form of the creator, and – further – it is a hymn to the temple: in its function of representing this divine-cosmic body of the creator. The temple itself is invoked, as the form of god.

In the following the sections of the hymnic sequence are summarized[154].

The god of Edfu is invoked by the names and epithets which we found hypostasized in the myth:

He is the god of *Bḥdt* and
he who is *on the High Seat (ḥrj St wrt),*
he is the *god of the mottled plumage (s3b šwt)*
who dwells in *Wṯst-Ḥr,*
he is *the Golden falcon (Bjk nbjt),*
the *Protector-of-his-father (Nḏ-jt.f),*
he is *He-of-the-powerful-face (Sḫm-ḥr),*
the *sgmḥ*-spear *who comes out of the water (pr m nwn),*
the *Winged Disk (Cpj),*
the *Ruler-of-the-Two-Lands (Ḥk3 t3wj),*
the *Lord of Msn (nb Msn),*

152 E I 13–20.
153 Alliot 1949 p. 154 ff.
154 Translation and commentary by Blackman and Fairman 1941 p. 426 ff. See also Alliot 1949 p. 151 ff.

the *Harpooner (j3wtj)* who pierces *the beast (nhs)* in
the *Place-of-piercing (St-wnp)*,
he is the *Strong-arm (Ṯm3-c)*,
he is the one *who runs swiftly (sjn gst)* and hurls
the *harpoon (mcb3)*,
he is the *ḏrtj*-falcon who slays his enemies,
he is the *gmḥsw*-falcon of *hidden forms (št3 ḫprw)*,
he is the one *who sends forth the flood at its time*
(bs nw r tr), the flood that sustains
the life of gods and men,
he is the one who is on the *hidden mound of Behdet*
(ḏw št3 n Bḥdt). (E I 14–15)

The gods who dwell together with the god of Edfu are invoked:

Hathor, his consort,
Harsomtus *(Ḥr-sm3-t3wj)*, the son of Hathor,
Khons,
Min,
Sokar,
Osiris,
Isis,
Tefnut,
Nephtys,
Nekhbit, and
the ancestors (nṯrw tpj-cw). (E I 15–16).

The different parts of the god's body are invoked. They are identified with cosmological phenomena as well as with gods:

His *living eyes (cnḫtj jrtj)*,
his *uninjured wḏ3t-eyes (wḏ3tj wḏ3tj)*,
his *eyebrows (smḏ)*,
his *nose (fnḏ)*,
his *lips (sptj)* which are identified with the two
doors of heaven, by which, when
opened, the earth lives,
his *tongue (ns)* which repeats (the words) and thus
renews life,
his *teeth (ṯswt)* which are identified with the
ennead[155].

155 Blackman and Fairman 1941 n. 80.

his *mouth cavity (ṯpḥt)* which is identified with
Mrt, goddess of music,
his *two* ⌜*collar bones*⌝ *(bjkṯj)* which are identified
with *the Two Kites (ḏrtj)*: Isis and
Nephtys[156],
his *two arms (m3wtj)* which are identified with his
two children, the two lions Shu and
Tefnut[157],
his *forearms (ḥswt)* which slay those who rebel
against him,
his *two hands (dtj)* which crush *the evildoer (nbd)*,
his *two wings (dm3tj)* by which he flies through
Nnt, the night sky,
his *belly (ẖt)* which is identified with the *sky*
(pt) with the stars (he himself
identified with the night sky – his
belly being equivalent to the
belly of *Nnt)*,
his *entrails (jmj-ẖt)* which are identified with
the ennead (in accordance with the
preceding image),
his *two body-supporters (tw3-ḥcw)* which are identified
with Isis and Nephtys who
support their father,
his two sets of *toes (s3ḥ)*[158] which are identified
with the two gods who are in peace
and harmony with his Majesty,
his *two feet (ẖndwj)* which reach the two skies,
his *nails (cnwt)* which are identified with the
living protecting snake-goddesses,
his *body (ḥcw)* which is identified with the gods
and goddesses who come into being
from Horus[a)] and who are beside him
(r gs.k: beside you)[159]. (E I 16)

a) *m Ḥr* can also mean "in Horus" and possibly both meanings are denoted here; the gods constitute his body and they are at his side as independent beings.

156 Wb V 597,3.4.

157 Wb II 403,10.

158 Wb IV 20,5.

159 Blackman and Fairman 1941 p. 423 n. 122.

The attributes of the god are invoked:

His *symbol of life (cnḫ)*,
his *w3s*-scepter,
his *nḫḫ*-whip,
his *ḥḳ3*-staff,
his Upper-Egyptian crown and his Lower-Egyptian crown with the *m3ct*-sign on it,
the emblem of *his two eyes (3ḫtj)*,
the *two feathers (šwtj)* on his head,
the *two supporting horns (cbwj)*,
the uraeus-snake on his head,
the *vulture-amulet (dm3t-pḏwt)* on the back of his head (E I 17).

The temple and its parts are invoked:

The noble *lion (m3j)* in front of the god, repelling the enemies,
His ancient *city (njwt)* of *Edfu (Ḏb3)*,
Wṯst which is identified with his mother Isis (in a wordplay on *wṯs:* uplift and *st:* seat and *3st:* Isis),
his throne in *Msn* which is identified with his *hidden (jmn) nest (sš)* in the Mansion-of-the-Falcon (the nest being the place where the falcon is hatched),
his *portal (t3j)* in the *Place-of-piercing (St-wnp)*,
his *gate (sbḫt)* in *St-wrt,*
his *hall (ḥ3jt)*,
his *pronaos (ḫntj)* for the god's appearance,
his other *palace (cḥ)* in *M33t-Ḥr*[160],
his ⌈...⌉[161] in the *Seat-of-the-two-gods (St-nṯrwj)*[162].

160 Wb II 10,6: "als Name des Gebietes von Edfu".

161 Blackman and Fairman translate: *thine inaccessible Naunet,* 1941, cf. note 151.

162 Name of the temple, E I 65; II 17.

the *chapels (ḥwt-nṯrw)* of the ennead,
the *mansion (ḥt-nṯr)* of *Msn,*
his *big hall Great-of-victory (Wr-nḫt),*
his *bḥdt*-throne in *St-wrt,*
his *sacred boat (wṯs-nfrw),*
his *images (cḫmw)* hewn on the walls,
all his *effigies (sḫmw)* of whatever name,
his *courtyards (wsḫwt),*
his *doors (sb3w),*
his *pillars (jwnw),*
and *all that is (wnnt)* of the temple. (E I 18)

a) It appears that the invocation is addressed to Horus of Edfu and to the gods and goddesses worshipped in his temple. The relationship between them is conceived along two lines: 1) the gods are the co-dwellers of Horus, and 2) they are members of his cosmic body[163]. The relationship is thus specified in accordance with the two metaphorical meanings of the temple.

It is the latter identification between Horus and Behdet and the gods which interests us most: the image of the cosmos-constituting gods formulates the theogenetic implication of the cosmogony from a henotheistic point of view. In this henotheistic conception of the cosmos, the hymn differs from the cosmogony myth with its narrative presentation of independently appearing gods, even if an underlying henotheistic concept can be discerned there, as well. In the hymn to the god of Edfu, the henotheistic concept of god is dominant; the many gods and goddesses are parts of his manifested being. The creator and the gods arise together, constituting an entity: the cosmos.

Again, it should be noted that cosmos is not conceived as a material body enlivened by a spiritual soul of god. The body consists of living gods – they are the cosmic phenomena equated with the body, the very components of cosmos. The body of the god of Edfu is the manifested divine being – not an empty form receiving god. Thus, the hymnic invocation of all gods of cosmos is an invocation of the god of the cosmos to unfold his being.

b) The invocation is also addressed to the temple of Edfu and to its parts and important properties. In this call is seen an iconological function of the temple. The temple is the cultic image of Horus and the gods constituting his being.

163 This conception of the relationship between the creator and the gods occurs frequently in Egyptian traditions, cf. e. g. the Leyden Hymn of Amun, IV No. 90 (In Junge's translation): *Die Neunheit ist zu Deinem Körper vereinigt. Dein Bild ist jeder Gott, der mit Deinem Körper vereint ist* (1978 p. 96).

It is the sacred icon of the creator; not merely a beautiful portrait but a living cultic body which can be addressed in a hymnic invocation.

Thus, in this morning hymn the body of god – that is, his manifested being – is seen under three headings: 1) a theological one – under which the body is the pantheon, 2) a cosmological one – under which the body is the cosmos, and 3) a cultic one – under which the body is the temple.

d) *The theogonic role of the king*

The king's relationship with the temple is a close one[164]. As we have seen, he is the master builder who conducts its foundation and construction; in doing so he puts into effect the command of Horus. Equally important is his position in the temple cult. The king is the cultic leader, being the highest representative of the Egyptian people. He is in turn represented in the actual cultic procedure by the officiating priest. As we have seen, the priest makes it clear to the god when he approaches the chapel that it is the king who has sent him[165]. In the reliefs depicting ritual situations, such as offering scenes, it is the king who performs the rite[166].

Under the theogenetic perspective, the building acts and the lithurgical acts of the king converge: the master builder creates the place in which god comes to life, and the priest creates the life led by the god and sustains it. We have noted the ritual embrace in the morning ritual, which transfers to the god his ka-life. In this act the king/priest represents the creator god[167]. The food offerings following upon the statue ritual are given by the king as lord of the fertile cosmos through the priest. In this act the life of god is maintained. Judging by the reliefs of the temple, this rite is central in the temple ritual. The most frequently occurring motif by far is the offering act. Offering scenes of varying kind are all over the temple, leaving no doubt as to the conclusion that this is the place where the gods receive their sustenance. The cosmological meaning of the offering is clearly expressed. On the base of the temple are depicted Nile figures and the fertility powers of the inundated nomes of Egypt, carrying their food while marching towards the offering-place headed by the king. They represent the life-sustaining capacity of Egypt, their products are

164 As regards the cosmos-sustaining functions of the king in temple-cultic context, see Derchain 1962 c.

165 See p. 97.

166 E X pls. CXIII–CXX; CXXXI–CLIV.

167 As regards this cultic role of the king-priest, see Hornung 1966 p. 24 ff.

presented as gifts to the god of the temple, the creator, and in this feature is implied a relationship of reciprocity between the cosmos and its creator: the god constitutes the cosmos, and the cosmos constitutes his life. In the reliefs above the base, the gods of cosmos are depicted receiving their food-offerings from the king who acts as a mediator administering the property of the creator. Similarly, on the pillars supporting the roof of the halls, gods are depicted receiving their sustenance from the king. The cosmological significance of the offering act is implicitly indicated by the metaphorical meaning of the pillars: they support the sky. In keeping with this imagery, offering scenes can be combined with scenes depicting the rite of *supporting-the-sky (tw3-pt)* in which the king is shown as the god Shu uplifting the sky[168]. In these reliefs, the temple as such is conceived as a vast offering-place: it reflects a cosmological offering-ideology.

A creative role of the king is, then, implied by the designation *lord of the rituals (nb jrjt jḫt)*[169]. He creates and maintains the divine life. Included in his theogonic functions is also that of uttering the words that call the god into being[170]. Consistently the cosmogony text is presented as a recital laid in the mouth of the king; it is introduced by the following words uttered by him: *I have come unto you, Falcon-of-the-mottled-plumage, Horus-of-Behdet, great god, lord of the sky, to bring to you my heart placed on its (right) place ...*

III. The iconological aspect of the temple

When addressed as a divine body, the temple moves out of the metaphorical sphere and acquires the function of a symbol. It is no longer merely an image of a dwelling or of cosmos; it represents the god of the dwelling and his co-dwellers, and it represents the god of cosmos and the pantheon existing in him. The temple is not only the place where the creator appears and in which he lives, but also the form of the living god. Being ritually approached as if it were a divine being, receiving hymnic invocations, the temple can be said to function as an icon of god. As has been stated previously, a significant decorative feature indicates this iconological function: when the temple appears in the light of the sun, the gods appear concurrently in the form of

168 Kurth 1975; Schenkel 1977 p. 19.
169 *Ad jrjt jḫt:* see Schott 1963 p. 103.
170 Assmann 1970.

the reliefs incised on the walls and pillars, literally filling every space of the abode, and in this way materializing the co-existence of the cosmos and its gods.

The iconological function of the temple affects all its parts including the decorative elements: they are constituents of a divine body and partake in god; so do even the hieroglyphic texts – the long texts carved into the building, a feature which the temple of Edfu shares with all Egyptian temples. A much-debated question has been what function these hieroglyphic inscriptions have. Different answers have been given[171] and from various points of view. None is satisfactory, and perhaps a single satisfactory answer will never be given. But one important point can be underlined: the answer should also be given with reference to the religious ideology of the temple. As we have seen, different meanings are ascribed to the temple with the help of its decoration. The landscape carved into it makes it the representation of chaos; the gods carved into it make it the representation of the cosmic pantheon. A similar relationship of representational identity can be found between the temple and the texts: they give their meaning to the temple; the building, so to speak, manifests their contents. The purpose of the inscribed texts is – when regarded from this point of view – analogous to that of the reliefs[172].

Within this perspective, the creation texts do not merely relate a creation story: they represent creation. Their being carved into the building is an act of defining the temple, in a similar way to that of the reliefs defining the temple, and the definition given by the creation texts is that the temple is the *cosmos that comes into being.* They are their own version of the creative words of the Heavy Flood which are *seen,* partaking in the theophany of the creator of the cosmos; they are meant to *appear.*

Another example of the representational function of the texts can be given. The building-text of the temple is carved in a broad band around the temple immediately above the base of its walls. The band has its own function within the layout of the walls. It serves as the decorative foundation for the

171 A usual explanation is that the engraved texts are a guarantee of duration – either for the sake of generations to come (e. g. Kees, who talks about an "Angst des Vergessens", 1977[3] p. 415) or for the sake of a prolonged effect of the rituals (e. g. Derchain 1962 c p. 63).

172 Within this functional context the texts and reliefs are of equal importance. On this point there has been a long discussion, however, alternating between views that stress the role of the reliefs at the expense of the text, e. g. de Rochemonteix, Rec. trav. 3 (1882) p. 85, and views that stress the role of the texts, e. g. Winter: "Ich sehe in den Beischriften den 'hörbaren' Niederschlag des geistigen Geschehens im Kultbetrieb, und in den Darstellungen nur Illustrationen dazu, d. h. aus künstlerischen Beweggründen zur Monumentalität gewachsene Vignetten zu den Texten", 1968, p. 16.

registers above the base; but in addition, it has a function within the total ideological conception of the building, as reflected in its reliefs. On the base are depicted the inundated nomes of Egypt carrying their produce. Above the building-texts are depicted scenes belonging to the cultic life of the gods of cosmos. The band of the building-text clearly represents the building of the temple above the inundated area, in which institution the life of the gods is maintained through the offerings of the produce from the fertile land.

In the application of the texts on the temple, an interest can thus be discerned in them as pictorial representations of concepts: the words not only denote the concepts but are *images* of them as well, and in this function they are on par with the reliefs.

I have endeavoured to show that when the temple is regarded as a religious phenomenon, all its decorative elements appear to constitute a unity, the elements being subordinated the comprehensive ideological framework[173]. It is to be expected that the above mentioned identification between the creator/his pantheon and the cosmos, which affects the meaning of all parts of the temple according to the morning hymn on the *St-wrt*-sanctuary, affects its hieroglyphic images as well; these *sacred carvings* belong to the manifested forms of god[174] – and within this frame-of-reference they partake in the *symbolic* aspect of the temple.

173 Important contributions to our knowledge of the reliefs and their relationship to each other as well as to the temple and its cult, are: Arnold 1962; Derchain 1962 b and 1966; Winter 1968; Helck 1976; Barguet 1980; Kurth 1983.

174 They could be called *b3w nṯrw*, or *b3w Rc*, Morenz 1960, p. 231.

C. The temple is a symbol of god

The material analyzed presents the relationship between the cosmos and its gods not only in terms of co-existence but even in terms of identification. This is particularly clearly stated by the morning-hymn inscribed on the façade of the *St-wrt*-sanctuary, which refers to the cosmos as the body of the creator[1] and to the gods he has created as the members of his body. By employing the image of the body, the relationship between creator and gods is also defined as one of (partial) identity.

The image of the divine body apparently lies behind the hymnic invocation to the *temple* and its components to come into being[2]. The link between the temple as dwelling of god and as itself the object of the cultic approach seems in this case to be the cosmos-imagery of the temple: it represents cosmos, and cosmos is the body of the creator. Thus the close relationship between cosmos and its creator which we have demonstrated in the preceding chapters, has been personified, and in a manner which allows of cultic approach – as it is a concrete object that appears as the cosmic body, the temple can function as icon of the creator.

One important inference from this temple-ideology with regard to its theological import should again be underlined: the creator is not thought to be *behind* the gods, as a kind of transcendent super-deity hiding behind the forms of the cosmic gods[3]. His *body* is his *being,* and the gods are parts of his being, they constitute him, cosmologically as well as cultically. The creator comprises them, but he does not transcend them ontologically; they are his components. In this context we should also recollect a characteristic of the concept of the cosmos presented earlier: this *body* of the creator is not a lifeless material shell which is filled with life by the divine spirit in the theogenesis, but from the very moment of its appearance a unit that can be qualified as living, divine reality. Similarly, when the temple is seen as the body of god, it is – from the moment of its appearance – living divine reality.

1 Identification between the creator and his work of creation is also witnessed in other Edfu hymns, e. g. E I 112; VIII 131.

2 The awakening is coming into manifested being, see p. 99 f. This equation between night-and-day and death-and-life is also documented in other Edfu texts, e. g. E III 50 (in Kurth's translation): *Sinke hinab in den Himmel als göttliche geflügelte Sonnenscheibe (cpj). Du sollst in der Frühe wieder zum Leben erwachen als der "Lebende" (sḫd m gbt m cpj nṯrj dw3.k r cnḫ m cnḫ)* (1975 p. 41).

3 See Hornung's presentation of earlier debates on the question of an Egyptian transcendent god, in *Der Eine und die Vielen* pp. 7–15.

I. The iconological character of the temple

In its aspect of divine body, the temple of Edfu is analogous to the statue of god in that it is an image of god which not merely describes him but represents him, actually and objectively. This is shown by the iconological functions of the temple.

a) On the definition of icon

The term *icon* is in this monograph used in its narrower sense to denote an image which is not merely a portrait of god but also thought to embody the divine presence. Thus, the content of the icon – namely the god that it re-presents – is not separated from the material icon: the icon is itself fused with the divine quality[4]; the icon of god is not a simile, but something of its own kind, namely, the cultic body of the god.

The iconological functions of the temple can be seen in rites directed to it and aiming at communication with the god represented by it, like those through which the Egyptians communicated with god in the statue. Between the temple and the creator of the world there is an iconological identification which implies that the temple can be the object of cultic attention in a fashion resembling that paid to the statue of god – and can be instrumental in the cultic communication. It receives hymnic addresses, prayers, and invocations; a mere simile does not have such cultic functions. This *portrait* of the world can function as a *symbol*[5] of the god of the world: it is the living, material body of the creator.

The icon has power to effect the presence of god. This function is well-documented in Egyptian religion and some aspects are known of the religious practises connected with it; that is, we are familiar with the anthropomorphic icon, the statue: it has the capacity to prophesy, to heal, to give various kinds

4 This definition is in conformity with the general terminology used within History of Religions, see e. g. E. Sharpe, *50 Key Words,* Richmond 1971 p. 28 f.

5 *Symbol* is in this monograph defined along generally accepted lines in History of Religions; it is a sign which not only *points to* something but *represents* it – and does so in a way which abolishes the division between that which it points to and that which is pointed to: "– the symbolic *presentation* is a coming-together, to the point of complete fusion, of the concrete and the spiritual", as Kahler has expressed it (1966 p. 70). In this feature the symbol differs from the metaphor – another representational image, which he characterizes as "para-phrase, parallelism, 'simile'" *(op. cit.* p. 72),

of assistance[6]. What this monograph intends to show is that also the cosmomorphic icon has power to effect the presence of god, and the capacity wielded by this icon is determined by the kind of god present in it: its capacity is to create the world and its gods. The effecting function of the temple is thus connected with its cosmographical and theographical aspects: the god of Edfu is conceived of as his work of creation, namely the cosmos and the gods. It is as a creator that he functions through the temple; when he is invoked, he appears as his work of creation.

b) *Remarks on the hermeneutical concepts applied to the material in the analysis*

The concept of *theophany*

In this monograph, *theophany* has been applied to define the divine aspect of cosmos. The definition given to the concept is *appearance of god.* Within the religious ideology that we are dealing with, this general concept will have for its theological content the transition of god from latent to realized being, and for its ontological content the transition from chaos to cosmos. We have traced a religious cosmology according to which the cosmic phenomena are by definition divine appearances: they are theophanies already from the moment of their coming into being – their theophanic character actually indicates their creation. Within a cosmology like this, the term *theophany* does not differentiate between *profane* and *sacred* cosmic phenomena, but between *latent* and *manifest* cosmic phenomena.

One of the questions that might be asked in connexion with this particular cosmology is whether it is valid only in a cultic context, or whether it reflects a general conception of being. We cannot exclude the possibility of the latter. This is not to say that we claim that the Egyptians lived in perpetual awe and reverence, but rather that their concepts of *god* and *divine* may have had a more everyday reference, and that they included these concepts in their general interpretation of cosmos as a matter of fact – but to the exclusion of a purely profane cosmos. Or to put it differently, that the polarization of *sacred* versus *profane* is not relevant in a general description of Egyptian interpretations of cosmos.

I do not attempt to answer this question, since it goes beyond the scope of this monograph, but I have wanted to indicate the kind of question that can arise from it.

6 Derchain 1962 a p. 187 notes 2,3,4,5.

The concept of *symbol*

Another word used in this monograph is *symbol.* It is applied when a cultic object is approached as if it were god, i. e. the word is used technically to denote this occurrence of a material phenomenon that represents god. There is no symbolistic theory of religion involved in the use of the term; a doctrine of a reality which transcends the natural and physical reality – the symbol pointing to this metaphysical reality through its allusions – does not seem to be the right ontological paradigm for this religion. The source material studied bears witness to a monistic ontology according to which *god* and *cosmos* are realities which are identified in a way which bars the interpretation that the cosmos is a vehicle of revelation for a trans-cosmic creator, or that the mundane phenomena are theophanies of trans-mundane divinities. There is a coalescence of appearing gods and appearing cosmos: the religious truth about the cosmos is that it is constituted by gods.

But there are, nevertheless, hermeneutical problems connected with the term symbol. By designating the temple as a *symbol of god,* we have explained the documented relationship of identity between god and object with a mediating auxiliary; and even with the formal definition we have given to *symbol* there is a great risk that the term will lead to a tendency to overlook the weight laid on the concrete object and to "spiritualize" it in a way which goes against the testimony of our sources. The way in which the term has been used generally associates to a dualistic ontology which distinguishes between god/spirit and cosmos/matter. In Egyptian thought god and cosmos, as well as spirit and matter, constitute a unit.

Because of this some historians have gone in the opposite direction when explaining Egyptian ontology, that is, employing a general theory of magic[7]. From one point of view, it offers a relevant perspective in so far as the theory of magic does not explain away the indissoluble unit of soul and body, spirit and matter, god and world, documented by the Egyptian material. If we use the designation magical to denote the formal structure of a concurrence of these complementary positions, the theory has its apparent merits; however, this concept also has its drawbacks. It has been used so often with a materialistic theory ("magic is a kind of scientific technology") that there is a risk of giving a reductionistic interpretation of this religion.

It seems that we have to conclude by saying that in our study of the Edfu temple we have come across a view of the world and a concept of god which are not satisfactorily rendered by any of these terms. This inadequacy should

7 First and foremost, E. W. Budge, cf. *Egyptian Magic,* London 1899, or *From Fetish to God in ancient Egypt,* London 1934.

be kept in mind when the terms occur in the following analysis of the iconological values of the temple.

c) *The iconological functions of the temple*

The iconological functions of the temple are analogous to those of the statue; but as the statue is anthropomorphic or theriomorphic it can be the object of rites which cannot be directed to the temple, such as the toilet-rites performed in the morning. Nevertheless, they have in common some important rites. Like the statue, the temple can receive *adorations* and *hymns* and *prayers.*

Examples can be found of adorations and hymns performed for the magnification of the temple during the dedication ceremonies. Dedicated to Re and Horus, the temple is praised in parallel adorations: under the name of High-Seat-of-Re, it is praised that its sovereignty be magnified; under the name of Seat-of-He-who-protects-his-father, it is praised that its splendour be sanctified. In both cases, the king performs the praisegiving:

Praising the High-Seat-of-Re since the primeval ones
with formulas of magnifying its sovereignty
(sw3š St-wrt-n-Rc ḏr sp tpjw
m 3ḫw nt swr šfjt.s)[8].

And:

Praising the Seat-of-He-who-protects-his-father
with the formulas of sanctifying its splendour
(sw3š St-nt-Nḏ-jt.f
m 3ḫw n ḏsr f3w.s)[9].

The king includes a description of the ritual situation in his formulas:

I have taken the roll and I read the hymns,
I elevate my voice to heaven,
I adore your temple with its secret names
(sw3š.j ht-nṯr.k m rnw.s št3w)[10].

When these words are uttered by the king, they leave their state of being *secret* and their content is revealed.

8 E II 34,4.
9 E II 63,8–9.
10 E II 63,11.

We have also seen an example of temple-adoration performed by *gods* during the *ḥts*-festival[11]. The temple is adored by Harakhte and by Tanen in a way similar to that of the king, the texts thus bearing witness to the mythological role played by the king-priest during the performance of the adoration. The praise starts with an apostrophe to the temple and continues with a recital of names conferred upon it – these constitute the *corpus* of the adoration. Harakhte raises his arm in the gesture of adoration and addresses the temple: *O, you place in which Apep was pierced!* ... The purpose of the praisegiving is the same as that of the praisegiving performed by the king: to magnify the sovereignty and splendour of the temple through the name-recital – the intention is declared in the words: *praising the High Seat with its beautiful names that exalt its might amidst the nomes (dw3 St-wrt m rnw.s nfrw sḳ3 b3w.s m sp3wt)*[12].

These are adorations offered to the temple personified, as if it were an anthropomorphic icon.

Hymnic appeals to the temple are not a phenomenon confined to Edfu: they are also documented in texts belonging to the temple of Kom Ombo. This temple, dedicated to Sobek and Horus, offers an example which we will quote because it both illustrates the personal aspect of the house of god, and also makes explicit the interdependency thought to exist between the temple and the cosmos. The hymn is recited by the officiating priest as he enters the sanctuary in the morning (in A. Gutbub's translation):

Formule pour entrer dans le temple d'Har(oëris), le Sobek seigneur d'Ombos, le prince des dieux, l'adorer (te temple), lorsqu'il y circule. A réciter. Salut à toi. Château du faucon ... (de) Haroëris seigneur d'Ombos, (et de) Sobek seigneur d'Ombos, tant que tu existeras, existera le ciel et vice versa, tant que tu existeras, existera la terre et vice versa, tant que tu existeras, existera l'horizon et vice versa; louange à toi, que l'acclamation et la jubilation t'entourent sur tout chemin, depuis que s'est installé le Ba de Chou avec le Ba de Geb, leur fils Osiris les accompagnant en toi, ils seront en paix (ou reposent) à côté de leur soeur, leur père Rê étant à leur suite, en joie ... district de la cessation du combat apelle-t-on ton nom.

11 p. 74 f.
12 E VI 319.

J'entre en toi, pur et lavé, je sors de toi doux et aimé; je suis Iounmoutef, qui purifie le Per-our; je suis venu pour accomplir le service divin en toi pour Haroëris seigneur d'Ombos, avec son ennéade divine, il donne toute vie, stabilité, force à son fils qu'il aime, le fils de Rê seigneur des diadèmes (Ptolémée vivant éternellement, aimé de Ptah et d'Isis) les dieux Philopators Philadelphes[13].

Several points pertaining to the icon-aspect of the temple can be noted in this hymn. First, there is the personalization of the temple inherent in the apostrophe: *praise (dw3) to you!*[14].

Second, there is a significant reciprocity between the temple and the cosmos; the priest states that insofar as the temple exists the cosmos will exist, and *vice versa.* This idea should be seen on the background of the iconological qualities of the cosmomorphic icon: the icon belongs to the creator. The words are more than a picturesque hyperbole: they refer to a creative capacity inherent in this icon. Behind them lies the idea that the world is dependent for its existence upon the temple and what goes on in it. More will be said about this idea in a later chapter. In connexion with this creative implication of the icon, notice should be taken of the cultic context of this hymn: it is recited in the morning – the cultic time of creation.

Finally, it might be noted that the *building*-aspect of the temple is retained; it conditions the characteristic trait of its iconological functions: the priest *enters* this cosmomorphic icon – and does so as a god, namely, *Pillar-of-his-mother (Jwn-mwt.f)*, a Horus-manifestation[15].

Like the statue, the temple also *receives prayers.* Ordinary people never entered the temple to see the statues of god. Even when a statue came out of the temple, carried in a litter, the people did not see it. It was hidden behind curtains. Nor did they see the interior of the building, since to move in this sacred world of gods was reserved for the priests. The people looked at the building from the outside, thus it was the temple exterior with which they had cultic relations. But even from the outside the temple appeared as a divine world. Its surrounding walls were covered with huge, divine figures: imposing, powerful, imbued with divine presence. Then there were the gates, the places of divine comings; through these the god appeared from his hidden state-of-being in the temple on festive occasions. The gates were the nearest places to god[16]. Thus the walls, and especially their gates, were approached in a cultic

13 Gutbub, *Kom Ombo* I (1973) p. 122 f.
14 Cf. Gutbub *op. cit.* p. 123 n. b–c.
15 Capart 1904 p. 88; Kees 1977^3 p. 336; E. Otto 1964, p. 70.
16 See Fairman 1954 p. 202 f.

way and became the recipients of prayers from ordinary people who never saw the icons of god in the temple interior[17]. On the southern gate of the temenos there runs an inscription that refers to this gate as *the standingplace of those who have and those who have not, in order to pray for life from the Lord of Life*[18].

d) *The temple undergoes the consecration ritual of the icon*

In order to function as an icon, the object has to undergo a special ritual. This ritual is called *Opening-the-mouth (wpj-r3)*[19]. The title presupposes an anthropomorphic icon, and the ritual is best documented when performed on statues. The bulk of the material comes from New Kingdom tombs. But any object that functions cultically as the icon of a god (or of a dead being) whether anthropomorphic or not, undergoes the ritual. The anthropomorphic image is not the only one to be found in the Egyptian iconography. Thus the ritual of Opening-the-mouth is also performed on sacred boats, scarabs, amulets[20]. Its application to such non-anthropomorphic objects is generally regarded as derived from its application to anthropomorphic objects (statues, mummies, coffins) and therefore supposed to have a general sacralizing purpose. In the case of the temple this would mean that the ritual should be regarded as a hallowing of the house for the purpose of setting it apart as a house not to be used by man. This is not an adequate understanding of the ritual. Its use on non-anthropomorphic objects may be secondary from a historical point of view, but it has nevertheless an iconizing function in these cases, as well. No change of purpose can be established with regard to the non-anthropomorphic objects on this point. As far as the sources go, the ritual has indiscriminately the effect of transforming the object into a symbol of god. When the ritual has been performed, the object can receive cultic attention. A temple, as well as other non-anthropomorphic objects, can be regarded as a "body" of god.

It is documented that the ritual was performed in Edfu on the statues of gods[21] as well as on the temple. The sources to the latter use of the ritual are two texts inscribed on either side of the exterior of the hypostyle hall,

17 Wildung in LÄ II 673.
18 E VIII 162,16–17; translation by Fairman 1954 p. 203.
19 E. Otto 1960.
20 RÄRG p. 487.
21 E I 173,3; III 277,3; 286,6; IV 242,15; V 238,10; VII 325,18.

on the friezes[22]. Blackman and Fairman noticed the importance of these texts as regards the temple ideology[23]. The two texts consist of enumerations of captions to the different rites of the ritual[24]. They both start by stating that the temple has been completed, and therewith introduce the object of the ritual. The text on the west wall enumerates the purification acts included in the ritual, and the presentations of oil, garments, collars for the statues, and the offerings of food gifts to which the gods are summoned and which are afterwards given to the craftsmen. It appears from these captions that the statues of the temple are also referred to. The text on the east wall is the more interesting one from our point of view, as it mentions the specific *opening* acts and also explicitly states that the temple as such underwent the ritual – not only its statues. The enumerated captions are as follows (in the translation of Blackman and Fairman):

To be spoken:
Wetjset-Ḥor of the Falcon of Gold, the temple of Rēᶜ,
it is complete with a girdle wall, the (very) spit
of Shu, fashioned by the Lusty Bull.*
Purification by the Lords of Purification.
Ptah takes his chisel to open the mouth
and Seker uncloses the eyes.
*Taking the sorcerer**.*
Presenting the finger of fine gold.
Proffering the Copper Adze of Anubis.
Ushering in the Courtiers: opening the eyes with
their adze and touching the mouth with the four slabs.
Beheading a smn-goose and
decapitating a goat.
Pointing at an Upper Egyptian male ox.
Slaughtering long-horned cattle and
strangling geese.
Presenting a great oblation of bread, flesh, and beer.
Opening the Mouth of Throne-of-the-Protector-of-his-
Father (wpj-r3 n Wṯst-Nḏ-n-jt.f).

* A designation of Ptah.

** A rod with the head of a ram at one end.

22 E IV 330–331.

23 Blackman and Fairman 1946 p. 75–91.

24 Cf. E. Otto 1960.

Censing its cult-chambers and
purifying its chapels.
Seker feeds the priesthood from the oblation:
gladdening their heart(s) with their largess.
Ceding Wetjset-Ḥor to its lord by His Majesty
(swḏ Wṯst-Ḥr n nb.s jn ḥm.f)[25].

The concluding act corresponds to the ceding of the statue to the god or the dead person in the statue-context[26]. The temple is given to its owner in a similar way as the statue is given to its owner when it has been transformed into his cultic body.

The actual performance of the ritual must be left to conjecture. Blackman and Fairman have put forward the following theory which might be a reasonable one: the priests have visited each hall and chapel, censing and asperging them, and performing the *opening* acts with the ritual instruments in mimetic gestures[27]. The ritual will have had the double reference to the different *parts* of the temple: rooms, reliefs, statues etc., as well as to the temple *as such*: seen as the total body comprising its parts. It is not unreasonable to draw into this ritual event – as Blackman and Fairman have done[28] – the anthropomorphic imagery of the temple presented in the morning-hymn of the *St-wrt*-sanctuary, and explain the *opening* acts as an iconization of the parts of the temple, by which they become members of the divine body of the god of *Wṯst-Ḥr*.

We have some knowledge about the time of the performance, but not enough. The texts document that the ritual was performed at the inauguration of the building, more precisely, at the inauguration of the hypostyle hall that took place under Ptolemy X Soter II[29]. In addition, the performance may have been repeated during the New-year festival *(wpt-rnpt)* as is suggested by Fairman[30]. This hypothesis seems a likely one. Thirdly, there may have been a daily performance, in an abbreviated version, during the morning ritual of the temple. Blackman and Fairman draw attention to the fact that the captions in the text on the western side of the temple enumerate rites which are also found in the morningly statue ritual, which is a toilet ritual comprising the use of water, natron, and oil as well as clothing, and the presentation of a meal.

25 Blackman and Fairman 1946 p. 77. The transliterations have been inserted by me.
26 E. Otto 1960 scenes 17, 18.
27 Blackman and Fairman 1946 p. 90.
28 Blackman and Fairman 1946 p. 84.
29 Roeder 1959 p. 321.
30 Fairman 1954 p. 187.

These rites are also found in the Opening-the-mouth-ritual, but the particular arrangement of the texts on the Edfu temple might suggest that they have been regarded as an independent class of rites – even though they can be incorporated in the *opening* ritual. The text on the western wall may be an abbreviated version of the daily morning-liturgy, the rites of which are united with the *opening*-acts in the Opening-the-mouth-ritual. It is not clear, however, what this analysis of the texts actually implies with regard to the question whether the *opening* ritual was performed every morning. The only indication of a regular morning performance seems to lie in the total mythological frame of reference of the morning rites: they refer to the "time of creation" and to the coming into existence of the *god* of creation. Thus the statue-ritual performed in the chapel of *Msn* includes the important embracing act – which is also found in the Opening-the-mouth-ritual[31] and through which the *ka*-life is transmitted to the icon. There is no reason why the latter ritual should not have been directed to the temple as well, as the idea of a daily creation is also connected with the temple, as we have seen.

e) *The temple-cultic* ka-*life of the god*

The ritual of Opening-the-mouth enables the object to function iconologically as the living body of god. The ritual itself states that the statue is made for the *ka* of the god or the dead man[32]. *Ka* is in this case synonymous with *life*[33]. The *life* of the god is affected by the rites directed to the icon. The same applies to the temple when it functions as the cultic body of the god: what goes on in the temple affects the *ka*-life of the god; the temple is made for the *ka* of the god[34].

Ka is a concept of life which has cosmical as well as genealogical references[35]. In addition, it includes the capacity to live[36], for which reason *ka* has also been defined *power of life*[37].

A comment shall be made on the relationship between the *ka* and the *ba*. *Ka* and *ba* are cognates, though they function slightly differently. The terms

31 E. Otto 1960 scene 54.
32 E. Otto 1960 scenes 65 B and 70 B.
33 Schweitzer 1956 p. 79.
34 Mariette, *Dendérah* IV 44a: The king has fashioned for the god "a monument for his ka"; Blackman and Fairman 1946 p. 85; Daumas 1951 p. 381.
35 Frankfort 1969[6] ch. 5; Schweitzer 1956 p. 74.
36 Schweitzer 1956 p. 71 f.
37 RÄRG p. 358.

are difficult to translate, because there are no equivalents in modern European languages. They both represent the entire self of the god (or man). But the *ka*-concept stresses the capacity to live, the vitality as such, while the *ba*-concept stresses the dynamic power of life.

When god is designated *ba*, he is seen as a dynamic being in the act of manifesting himself – the god who comes into being. This *ba*-power can manifest itself as animate and inanimate objects, as well as other gods. In the cosmogony the *ba* of Horus appears as the Winged Disk *(Cpj)*: it is the Flying Ba *(B3-ḥdd)*.

This feature points to another aspect of *ba: ba* links the god with his own pre-being. It is the *ba* that effects the transition from the underworld *(d3t)*. Structurally, *ba* therefore corresponds to the transcendent *soul* of dualistic systems, and this is why the word sometimes has been translated "soul". However, the translation is not satisfactory because *ba* is not equivalent to soul in modern European sense. Egyptian anthropology conceives of *god* (and *man*) as a unit of faculties that can be classified as psychical and physical[38]. The dualistic paradigm of a being constituted by the complementary soul and body has no place in Egyptian thought. Thus, *ba* (and *ka* – which is also sometimes translated "soul") refers to the entire personality, and "person" or "self" might in many cases be the nearest equivalents to the terms[39]. The *ba* can be invisible (namely, in the underworld *(d3t)* where it is not seen) or manifested (the perceptible, cosmic phenomenon), but this distinction does not coincide with the categories of soul and body. We have seen that when the *ba* of Horus comes out of the underworld to *unite with* the temple, he comes as the Winged Disk *(Cpj)*[40], and the act of *uniting* implies that this *ba*-quality is transferred to the temple so that it receives the power to come into being and manifest itself. It might be added parenthically that in mortuary context *ba* connects the dead person with *d3t*; when the dead person comes out from his tomb in the day, it is his *ba* that comes[41]. *Ba* thus implies ontological transition, and refers more to a state, or mode of existence, than to a permanent quality.

38 Žabkar, LÄ 590, and 1968. As Žabkar's studies have shown, there was no identification between *ba* and *soul* even in the latest period of the Egyptian history: "so foreign was the idea of immateriality or spirituality to the concept of the Ba that the Christianized Egyptians found the word *b3* inadequate to express the Christian idea of soul and borrowed the Greek word 'psyche'" (1968 p. 162).

39 Žabkar 1981 p. 154 n. 99.

40 See p. 111 f. Cf. Assmann 1969 p. 193.

41 See p. 110. *Ad* the sexual overtones of *ba* as regenerating life, see Bergman 1970 p. 86 ff.

Ka refers to the life-quality as such – the inherent, permanent capacity to live possessed by every created being. As long as god lives he is, or has, his *ka*. His *ka* can appear in different forms, though, and all of them belong intrinsically to the qualification of god as a living god. *Ba,* on the other hand, designates the position occupied by god when he effects the transition into life. Thus god is *ba* when he comes to the temple to *unite with* it – while his *ka* is associated with the temple when it is thought of as the god's residence[42]. It is the *ba* that brings the temple into appearance – while it is the *ka* who sits on its throne. This relationship between the *ba*-nature and the *ka*-nature of the god in relation to the temple is brought out in a dedication text which we will quote for the purpose of illustration (in de Wit's translation and transliteration):

Le noble disque ailé se montre dans le ciel en tant que Celui-de-Behedet, le grand dieu, seigneur de ciel. Il parvient à son nome, il s'unit à sa chapelle, son ka s'assied sur son trône et il contemple ce magnifique ouvrage qui a été fait par son héritier bien-aimé, le roi de Haute et de Basse Égypte, Ptolémée X.

cpy špś di.f tp.f m nnt m bḥdty nṯr c3 nb pt śpr.n.f śp3t.f im3.n.f śtit.f śnḏm k3.f ḥr mn-bit.f dg3.n.f k3t.ṯn nfrt irw.n iwc.f mry.f nsw bit iwc nṯr mnḫ nṯrt mnḫt rct śtp n ptḥ ir-m3ct-rc[43].

Compare the following presentation of the temple: *The great seat of his ka in which he appears and rests (p wr n k3.f ḫc.f ḥtp.f)*[44]. The residing-aspect is focused on, and the god is seen as his *ka*. When the *ka* is presented as sitting on the throne of the temple, there seems to be an amalgamation between the terms *ka:* life-capacity, and *ka:* the person of the majestic god[45]. The latter meaning of *ka* is prevalent in the following example, where the temple is conceived of as *bw pn ḫrp k3 jm*[46], which can be paraphrased: "this place where the god-king reigns".

The divine life lived in the temple encompasses all main categories given to the *ka*-life in Egyptian traditions. It is lived *anthropomorphically* in the statue, the temple functioning as the dwelling in which god lives like man, sleeping, awakening, eating, drinking, displaying all those activities which are represented by the rites surrounding the statue. Further, the *ka*-life is lived

42 This is also valid when the icon is the statue; the god comes as *ba,* and his *ka* resides in the statue. For comparison with Denderah-traditions, see Daumas 1951 p. 394, cf. p. 398.

43 E VII 3, 1 f.; translated by de Wit, CdE 72 (1961).

44 E VII 1,13.

45 See p. 83.

46 E VI 176,4 f.

theriomorphically – in the falcon statues, and also in the living falcon kept on the temple premises[47]. Finally, it is lived *cosmomorphically* in the temple when the temple functions as icon, the *ka*-body of the cosmic god.

The ritual creation of this *ka*-life is effected along two main ways:

1) When the icon is a statue, a ritual transference of *ka*-power is performed through an embrace; the priest represents the theogon in this act. When the god thereby receives his *ka,* he is born. The creative embrace has an important place in traditional Egyptian theogony. The rite takes place during the daily morning-ritual of the statue.

2) When the icon is the temple, its life is created through invocation. As we have seen, the names of god recited in the morning-hymn have creative functions: they *magnify* the god according to the names and epithets. When names and epithets are bestowed upon the temple in the hymnic invocation, they similarly imply a creation of the life of the temple.

As regards the life-potential inherent in the name, a feature of the creative procedure adopted by the long cosmogony text can be considered with interest: more precisely, the special formula applied when the mythical landscape is named. The divine life of cosmos is reflected in the mythical landscape – which is *named* after mythical events: the names given to this landscape draw their content from the events. These are all *k3*-names – while the names given to cosmos as a geographical entity are *rn*-names. Thus there seems to be a play on the words *k3: name* and *k3: life* in the presentation of the coming-into-being of the divine landscape. We will give some examples:

The name (k3) Water-of-fighting has its content from the mythical fighting of the protector.

The name (k3) Floater has its content from the mythical floating reed.

The name (k3) Support-of-Horus has its content from the mythical settling of the Falcon.

The name (k3) Flying Disk has its content from the mythical event of the Lord-of-flying coming to do homage to the god of Edfu.

And so forth.

Together, these *k3-names* present the divine *k3-life* as it unfolds itself in cosmos and in the temple.

47 Fairman 1954 p. 189 ff., and 1960 p. 80.

f) Personification of the ka-life of the temple

The life-imparting capacity of the temple can be personified as a mother goddess. The feminine characteristic of the personification is partly due to a philological cause, as there is a correspondence of gender between the personification and the designations of the temple: *ḥt (mansion)* – or the more frequent *st (seat)*. But the feminine gender is not merely a matter of philology, it is also conditional upon a mythology of the life-conveying capacity of the temple, according to which the theogenetic function of the temple is seen in terms of the mother who bears her son. In Edfu, the mother who personifies the temple life is Isis – the traditional mother of Horus, not only in this town. The mythological relationship between the temple and Isis is, nevertheless, complicated, because it draws upon different sources. We shall not make it our concern to trace the historical development of Isis as the personified life-giving capacity of the temple, only limit ourselves to singling out moments that are of significance to the explanation of the phenomenon from our temple-cultic point-of-view.

1: An ideology of the *throne, the high seat (st wrt)*[48] of the king of Egypt and his mythological correlate, Horus, ruler of the cosmos, is involved. Isis is intimately connected with this kingly *high seat.* Formally, the connexion can be seen in plays on her name, such as *3st: Isis – st: seat*[49], and in the emblem on her head: a high seat. Thus, when the king is seated on the throne, he is associated with the mythical image of Horus seated on the lap of his mother.

2: This seat of the mother was early equated with the *cosmological* high seat *(st wrt)* of the sungod creator – the place where the theogony commences. Already the Pyramid texts refer to *the high seat which creates the gods (st wrt jrt nṯrw)*[50].

3: The *st wrt* has a cultic correlate in the temple, and is here the place with which the ba of the sungod *unites* to create the gods of the cosmos. In this function it can be said to correspond to the *st wrt jrt nṯrw* of the Pyramid texts. The throne-value is, however, also found with the temple-cultic seat. As we have seen, the ka of Horus *reigns* in this place.

48 Bergman 1968 p. 123 ff.

49 There appears to be no etymological connexion between *3st* and *st,* see Kuhlmann 1977 p. 96 ff.

50 Pyr.t 1153–1154; cf. Bergman 1968 pp. 127, 133, n. 3.

4: In the apostrophes and invocations to the personified temple of Edfu, the temple is addressed as *St* or *St-wrt,* or as *Wṯst* – the local version of the cosmological first seat of the god; *Wṯst* could just as well be translated *Seat* because of this equation. These are the names of the temple preferred when it is personified. They all associate it with Isis, through the mythology of the seat. In the text to the coronation of the falcon, Isis is explicitly identified with the temple under the name *Wṯst-Ḥr*[51].

5: Isis is the traditional mother of Horus[52]. In the Edfu traditions this mythologem is connected with *Wṯst* through wordplay on the following Isis-Horus-constellation, which adds another meaning of *Wṯst* to the ones we have noted in the cosmogony texts: *Isis lifted up her son in Wṯst, and Support-of-Horus became its true name (wṯs 3st s3.s ḫnt Wṯst ḫpr Wṯst-Ḥr m rn.s m3c)*[53]. Thus not only is Horus the king-god and the creator-god brought to this temple-seat of Isis, but also Horus the son of Isis, Harsiese.

6: This mother of Horus has been identified with the cosmological mother, the goddess of heaven who gives birth to the sungod – a mythologem which is central in the morning ritual and also in the cosmological temple-imagery. When the god is in the dark temple, he is in *Nwt,* or in her *d3t*-variant: *Nnt;* when he leaves the darkness, he appears as the sun, i. e. he is *born*[54]. Through this identification with the goddess of heaven Isis is associated with the temple not only as *Seat* but also as *Heaven* – the other cosmological source of divine origin located within the temple.

All these threads are woven together into the many-coloured tapestry typical of Egyptian mythological thinking.

I shall give a few examples of temple-personifications which allude to Isis. In the morning-hymn on the *St-wrt*-sanctuary, the temple, under the name of *Wṯst,* is identified with Isis, in her role as mother of Horus (in Fairman's translation): *Wetjest which lifteth thee skyward, thy mother Isis protecting thee*[55]. Both from a literary and a temple-architectural point of view, this identification between the temple and the mother is well chosen. For its cultic purpose, the hymn has the creation of the god's ka-life, and the *St-wrt*-sanctuary contains the cosmological seat which is associated with the theogenesis – it is the seat with which the ba of the creator unites when the doors are opened, the time when the hymn is recited.

51 See p. 140.
52 Bergman 1968 p. 134 f.
53 E VII 10,6.
54 See p. 100 with note 70; cf. Daumas 1951.
55 Blackman and Fairman 1941 G III 7.

Another example is found on the inner face of the enclosure wall, in the scenes pertaining to the coronation of the *living Falcon*[56]. The relief in question presents a scene where the falcon and Horus of Behdet are grouped together with the personification of the temple; the falcon is on his *srḫ*-throne, placed on a pedestal[57]; he is given the epithet *Bjk c3 ḥrj srḫ, the great Falcon on srḫ*[58]. Behind him Horus of Behdet is seated, also on a pedestal. Behind this seated Horus are his mother, *Wṯst-Ḥr,* and his nurse. The name of *Wṯst-Ḥr* is written on the head of the goddess. The caption to the figure says: *Words to be spoken by Wṯst-Ḥr, the Powerful-one among the towns, the Great Seat who protects her son (ḏd mdw jn Wṯst-Ḥr Wsrt ḫnt njwt St-wrt ḫw s3.s)*[59]. From the last words it appears that this personified *Wṯst-Ḥr* is identified with Isis. As regards the appellation *Wsrt,* it occurs as an Isis-appellation in Greek time[60].

There are similar personifications from the New Kingdom of the theogenetic capacity of the temple, which show that the phenomenon belongs to an old tradition[61]. The New Kingdom-examples shed light on the Ptolemaic ones. They are found in mortuary temples at Thebes – more precisely, in the temples of Ramses III at Medinet Habu and of Seti I in Qurnah[62].

In Medinet Habu, there is a relief that depicts the temple as a goddess embracing the king[63], a gesture which expresses both transference of life[64] and identification[65]. The relief can be linked with words spoken by Tanen to the temple of Edfu on the festival of *ḥts-ḥb*. When Re Harakhte, Tanen, Thoth, and Seshat arrive at the temple, Tanen calls upon the temple to open her arms and receive the god: *open your arms to receive His Majesty Harakhte!*[66]. These words reflect the mythologem of the mother who receives her son – so that he unites with the source of his ka-life. The mythologem may be combined with the notion of a return to the mother of the night sky, the temple being conceived according to its *d3t*-value: when the god enters the temple he returns to the

56 E X pl. CLIV = E VI 298–304.
57 This throne presents the god as owner and inhabitant of the temple, see Kuhlmann 1977 p. 85.
58 E VI 303,15.
59 E VI 304,8.
60 Bergman 1968 p. 266.
61 Actually, the tradition has roots in the Old Kingdom, see Wilke 1934.
62 See Nelson 1942.
62 Nelson 1942 p. 130 pl. IV.
64 See pp. 100 f., 120, 137.
65 As a rite of identification it is represented in the tomb of Tutankhamon, where the king is depicted embracing his Osiris-form; Piankoff 1955 pl. 7. See also Assmann 1969 p. 103.
66 E VI 319,11.

latent life within his mother from which he will re-appear as the morning-sun, newborn. In the address to the temple, given by Re Harakhte on the same occasion, this idea is hinted at. The sungod addresses his temple as *St-wrt* and qualifies it as the place in which he has *hidden (jmn)* himself in order to make his *appearance (ḫcj),* a word connoting the sunrise[67].

At Qurnah there are two reliefs of interest. In these the dead king, Osiris-Seti, is depicted seated, while behind him the ka of the temple stands personified as his mother. The ka-sign is on her head, indicating her affinity with his ka-life, and within this sign is written her name – which is that of the dead king's temple (in Nelson's translation): *The temple (called) "Seti-Merenptah-is-Glorius" in the estate of Amon on the West of Thebes.* The temple says (in Nelson's translation): *Behold, I am behind thee. I am thy Temple, thy mother, forever and ever*[68]. The relationship between the dead king and his temple is here shown as one between the son and his mother. The temple holds his ka-life which is given to him: the vital power given anew to the dead king, Osiris-Seti, god of latent life. The temple is the source of his ka-life; it is the icon which effects his coming into being again. The transference of the ka-life from his mother is indicated by a gesture of imposition: the mother places her hands on the shoulders of the king – an analogy to the embrace. At this point the position of the goddess is to all appearances determined by her underlying association with the seat – even if this is not expressed in any other way than by posting her behind the seat, the traditional place of the goddess of the seat. With this act of imposition the king unites with his own ka-life.

The kind of life given to the dead king by his temple is the cosmologically defined life which is represented by the temple of Edfu; the king receives the same life which is given to god. This fact is conditioned by the monistic ontology of Egyptian thought, and the idea that the king takes part in all-cosmic life in the same way as the god of cosmos can be found in the earliest textual sources to Egyptian religion[69].

The god and his icon constitute a unity: the ka-life of the god. But in the temple-mythology outlined above, this unity is give an internal dialectic structure. With the help of the ka-personification, the god is related to his own temple-life as if it were a person; the ka-life of the temple is his mother. The mythology of the mother and the son presents the relationship between the god and his life as a development. This mythological conception of his life stands beside the one previously noted: that the creator is his own creator;

67 Wb III 239.
68 Nelson 1942, p. 131; fig. 5.
69 In the Pyramid texts, e. g. 508, 509, 523, 539.

he himself creates his ka-life by *uniting* with his icon, the *Seat;* as the ba of the sungod he is his own power to come into life; his life is self-generating. The two different conceptions of the generating of the life can, nevertheless, be combined in the Egyptian mythological vocabulary. Thus Amon of Karnak is called *Bull-of-his-mother,* i. e. Begetter-of-his-mother *(K3-mwt.f)*[70]. When this name is applied to the solar god, the mythologem associated is that of the god who makes his mother pregnant by going into her mouth, and who comes out of her as her son[71].

A pair-relationship between the creator and his temple is also reflected in the Kom Ombo-hymn quoted above. When the priest walks into the temple he presents himself as *Pillar-of-his-mother (Jwn-mwt.f)* – possibly with an implied reference to the mother as sky-goddess[72]; the priest .exercises his creative function in the role of the cosmos-sustaining Shu[73]; but the name also connotes procreation[74].

Through this mythological structuring of the life of the temple-god, a dynamic movement within the godhead can be described. In these images of the triad father-mother-son, the Egyptians developed their ideas concerning the theologumenon of the auto-genetic god.

II. Distinguishing features of the icon as regards cognitive content and cultic functions

A characterization of the concept of god presented by the temple

a) *The temple is the icon of a creator immanent in his work of creation*

An equation between the creator and the cosmos is documented in the Edfu sources. This is no god transcending his work of creation – the cosmos is his

70 The designation *k3: Bull* is conditioned by the identity of the mother: she is the skygoddess in her mythological apparition of a cow.
71 Frankfort 1969⁶ p. 180; Assmann 1969 p. 317.
72 RÄRG p. 325; cf. pls. 76 and 77 on p. 303.
73 It is documented in E III 78,11–12 that the priest presents himself as Shu as he enters the temple.
74 Frankfort 1969⁶ p. 169 f.; Assmann 1969 p. 130; Daumas 1951 p. 376.

body[75]. For a closer definition of the creator two questions should be considered: 1) what kind of cosmos is involved in this concept of god, and 2) what kind of immanence is involved?

ad 1: The ontological identity of the god of the temple

There is one significant feature appearing from our analysis of the theophanic and cosmophanic aspects of the temple, which should be emphasized because of its implications with regard to the concept of god referred to. The cosmos equated with the god of Edfu is identified with the town of Edfu. In the morning hymn of the *St-wrt*-sanctuary, the cosmic creator is invoked as *the god of Bḥdt,* the god who dwells in Wṯst-Ḥr, and the one *who is on the hidden mound of Bḥdt;* his *ancient town of Ḏb3* is invoked together with him. As we have seen, the area of *Bḥdt, Wṯst-Ḥr* and *Ḏb3* is the cosmos of the cosmogony texts. This geographical area is topographically specified as the mounds upon which the dwellings of man are built, and the inundated soil which is cultivated by man through irrigation agriculture. It is, in other words, man-shaped nature, nature formed and controlled – not in opposition to culture but integrated in it. This is the kind of cosmos which belongs to the god of Edfu and with which he is equated. The ontological identity of the god of this town is – this town[76].

Much has been written about the relationship between Egyptian gods and nature. H. Frankfort is among those who have given us the finest insights. The Edfu sources confirm the general intrinsic closeness between gods and nature demonstrated by Frankfort, but also, they make it clear that the nature with which the creator of Edfu is concerned, is formed by the Edfu culture: it is the nature which is made into cosmos with the help of man.

75 Our study has confirmed E. Hornung's conclusion from his general study on the Egyptian gods: "von jener eigentlichen Transzendenz, in der sie sich über Raum, Zeit und Schicksal hinwegsetzen und ihr Wesen ins Absolute, Schrankenlose weiten, kann in Ägypten nicht gesprochen werden" (1973 p. 186).

76 As I have mentioned earlier, the cosmos-values of the temple also comprise that of *Egypt*. The relationship between these two meanings of cosmos, *Edfu* and *Egypt,* should be worked out in a special study which takes into account the different cultic contexts involved, and also the different appearances of Horus: for instance, when Horus appears as *king of Egypt,* e. g. during the *sed*-festival, the corresponding cosmological value of the temple is *Egypt* (Frankfort 1969[6] p. 86 f.). I have focused on its value of *Edfu* – the prominent identity bestowed upon it by its nomenclature and texts of creation.

ad 2: What kind of immanence is ascribed to the god of the temple?

Perhaps the most apt characterization of the relationship between the creator and the cosmos is that it is a relationship between god and his manifested life: the cosmos is his living being. Cosmos is not a medium of appearance; the immanence of this god is no revelatory visit to the world of man from a transcendent world of his own. This point is an important one and it deserves attention. There has been a tendency[77] in discussions on the immanence of the Egyptian gods to regard *god* as a phenomenon abstracted from his cosmological form – to separate *divine being* from its *manifestation,* and thereby to assume, implicitly or explicitly, that *god* has independent existence in a world outside the world of man and from which he appears by revealing himself in cosmic objects[78]. But the life of the god of Edfu lies embedded in cosmos, cosmos is *the god that lives.* The god we are dealing with *is* his manifestation, the latter being his *modus* as living god. The term *revelation* can therefore be a misleading one in a description of Egyptian religion, even if it can be used – provided the revelation is understood as the god-who-has-come-into-being. This is no religion of incarnate god; the god of Edfu is not god in cosmic guise: he constitutes cosmos – or he lives not, in which case there is no cosmos.

It follows that this relationship between god and cosmos does not find analogies in religions like the Jewish or the Christian ones, where the cosmic phenomena can safely be said to be the media of revelation for a transcendent god. The life of the Egyptian creator is cosmic; he does not transcend the phenomenal world, but is immanent in the sense that he constitutes it.

The only form of *transcendence* that takes place within this monistic system, is the movement into *latent cosmic being: chaos*[79]. With reference to this content definition of *transcendence* one might say that the Egyptian *revelation* corresponds to the *coming into manifested being* – as opposed to latent being. The concepts of *transcendence* and *immanence* can thus be used in the formal analysis of the cyclic coming into being of god; they can be justified as terms designating the *hidden* and *manifested* life respectively – referring to phases of a repeated coming into being; only, god does not appear from another world, but from his own potential life. The dynamics behind this self-generating

77 Among the exceptions, Derchain with his naturalistic interpretation of *god* appears to be most consistent, although his outlook is somewhat reductionistic, as his concept of *nature* is based on a modern Western definition, see p. 153.

78 See Hornung 1973 pp. 7–17.

79 Cf. Assmann 1975 b p. 24 f.

process is expressed in the mortuary mythology in terms of a union between Re and Osiris[80].

This innate cosmic character might be designated the backbone of *the* Egyptian concept of god – as it is to be witnessed in different sources from different places. It is an old idea, already found in the Pyramid Texts – which are noted for their lack of transcendent gods – and it is interesting that its continuous influence can be traced down to Hellenistic times. As we have pointed our earlier, this immanence of god cannot be characterized as dead material enlivened by divine spirit. The cosmos is the *god that lives,* god is thus spiritual-and-material. Correspondingly, the god whose presence is incorporated by the temple is no transcendent spirit captured within his material icon; the icon is living, material god.

Behind the texts and ritual acts of the Edfu temple there can be discerned a theocentric ontology, the concern of which lies with the innermost nature of the world. It states that man's world is basically divine being, and even material phenomena are divine being: the inundation, the sun, the mounds, the irrigated land, the town. There is nothing in man's world that is not a form of god; there exists no "mere" material. In the presentation of this cosmos there is an accent on the divine side of life: it all starts with god. Accordingly, the Edfu cosmogony gives priority to the appearance of the gods: with their appearances the cosmic places have come into being. Or, as the priority has been formulated in the Bremner Rhind papyrus: *When I came into being, 'Being' came into being*[81]*: Being* is subordinated to *divine being.*

In order to bring into relief the outstanding features of this immanent god, he might be contrasted with the utterly transcendent god of certain Gnostic systems in which god has nothing whatsoever to do with cosmos. He may be its ultimate and indirect source, but he does not partake in its nature or in its functions. On this point the ancient Egyptian national religion and these Hellenistic religions – also found on Egyptian soil – represent diametrically opposed theological systems.

b) The temple formulates a henotheistic concept of god

The temple is the icon of the creator in whom the gods have their being. This concept of god is *mythologically* expressed in the morning hymn inscribed

80 E. g. Dondelinger 1973 p. 103.
81 See p. 92.

on the *St-wrt*-sanctuary, which conceives of the temple as the body of Horus and the parts of the temple as the gods constituting the cosmic pantheon. The *temple-architectural* expression of the theogon who comprises the presence of the other gods is found in the reliefs of the gods engraved on this icon of their creator; the effect of this device is that the theogon appears as the pantheon he has created. Through this icon, then, a cultic meeting with the plurality of the creator can take place. The temple is the icon of Horus who has differentiated himself as theogon. This aspect of the temple as it appears with all its reliefs of gods agrees with the apt characterization of the Egyptian creator given by E. Hornung on the basis of mythological evidence: "Der Eine und Undifferenzierte des Anfangs hat sich durch sein Schöpfungswerk selber differenziert, 'zu Millionen gemacht', der Mensch kann ihm nur in der Vielheit der geschaffenen, vergänglichen und wandelbaren Götter begegnen"[82].

This concept of god can best be called *henotheistic* as E. Hornung has suggested, applying a term which accentuates that the one god manifests himself as many gods. *Polytheism* thus belongs intrinsically to this concept of god[83], as distinct from *monotheism* which denotes belief in one god and dispenses with the many gods. The henotheistic belief does not conceive of the many gods as less real or less important than the encompassing god; indeed, they are cultically requisite in the temple of the creator (all temples of Egypt had cults of other gods beside that of the god of the temple, the creator). One of the characteristic traits of the temple cult is that it encompasses the cults of the many gods. They all constitute the cult of the temple. Thus, when Horus is approached as creator in the morning cult, the ritual is carried out for all gods in the temple[84]: The coming-into-being of the creator implies the coming-into-being of the gods. This total cultic situation is, then, the *temple-ritual* expression of the concept of the henotheistic god.

Henotheism and monotheism are correlated with different ontologies. While henotheism is accompanied by a cosmos in which gods are immanent – in some form or other – the monotheistic god may have created cosmos but he lives above it and is not dependent upon it for his being. He may, though, intervene in the affairs of the world through incarnations. Because of these ontological distinctions the temple cannot function as the personified icon of god in a monotheistic system, even if it can represent the work of creation in its aspect of witnessing god's function as pantokrator.

82 Hornung 1973 p. 179.

83 Hornung has shown this with regard to the Egyptian gods in *Der Eine und die Vielen. Ad* the discussion concerning *monotheism* in Egyptian religion, see *ibid.* pp. 1–19.

84 Fairman 1954 p. 180.

Pantheism is another term denoting a cosmos-immanent god, which has been applied to Egyptian religion[85]. The term refers to an all-pervading deity, a god that permeates the world as a kind of divine essence or principle. The concept is basically impersonal and resists concretization. Because of this feature it is not conformable to the god witnessed by our material, even if pantheism may be combined with polytheism on the cultic level, which provides the necessary limitation that confines the god in forms permitting cultic communication. But the many gods are in such cases theologically subordinated the impersonal and boundless god – they are less “absolute” and often thought to be the forms of god suitable for the masses, while the wiser people have discovered their true although illusive character[86]. The pantheistic god is not really cultically relevant, for to be so would require a certain degree of limitation; the pantheistic god has no cultic body as has the henotheistic Egyptian *pantheos* – who is cosmic and personal and limited[87].

c) *The temple is the icon of a dynamic god*

The temple presents the creator as a *dynamic* god. He does not repose statically in a world of no change as a kind of eternal and invariable constant behind the changing, created world. The creator is not a god from eternity, uncreated himself. On the contrary, he comes into being with and through his own creation. The temple shows him precisely as this dynamic god – continually coming into being. Metamorphosis is the immanent creator's mode of creating, and his icon shows his metamorphoses. The temple is itself a dynamic entity as it is an icon which reflects the transition from dark chaos to illuminated cosmos. It is the underworld of the cosmos and the place of the hidden life of the creator, where he awaits his coming out; it is his place of appearance, displaying his *ḫcj*. As the statue shows the anthropomorphical cycle of being which the god shares with all human beings – a cycle comprising sleep and awakening, death and rebirth, the temple shows god's cosmological cycle of being. It is the image of the creator who comes into being by transforming himself into cosmos, and who takes part in the rhythmic movements between latent and manifested being, not elevated above the world but existing in its very texture and development.

85 Breasted 1912.
86 Cf. Hinduism.
87 See below.

d) *The temple formulates god as a cultic phenomenon*

The cultic functions of the temple belong to the factors which define the concept of god of this icon. Thus a basic qualification of the god lies in the fact that he is god of *a place of cultus,* that is, where gods are invoked and the creator magnified with all the names and offerings which make him lord of the world. The life imparted to him by this icon is cultic life.

All icons represent god as a cultic phenomenon insofar as god becomes cultically present through the icon. What characterizes the temple under this aspect is that it works the cultic presence of the all-encompassing creator of cosmos. When Horus is adored in one of his anthropomorphic forms, only part of him is represented. Through the temple, on the other hand, man relates himself to the god of the totality. Another characteristic feature of this representation of god is its cultic function of creating the cosmos and its gods: through the temple the entire cosmological pantheon comes into being. This theogonic capacity makes the temple a prerequisite in Egyptian religion. The temple is more than a picturesque account of the cosmological identity of the creator and his gods; this vast institution, this expensive household is not upheld to illustrate in a spectacular way that cosmos is constituted by gods. It is the place where the gods come into being.

The dependency of both the gods and the cosmos on the temple is also documented in other Egyptian sources. A standard way of expressing it is to say that because of the temple-cult the sky will not fall down on the earth, and the sun will follow its course, and the Nile will not dry up[88]. These statements are not rhetorical hyperboles for the sake of artistry. They refer to an interaction between temple-cult and cosmogony, and practical consequences are involved. There is an interdependence between the cultic presence of the gods and the factual life of the cosmos, which necessitates that temples are built and cults instituted – otherwise the land of Egypt will disintegrate. As such, the temple and its gods are presented on the so-called Restoration-stela of Tutankhamon, from which we will quote because it illustrates the relationship between the temple-cult and the state of cosmos.

The stela was found in the temple of Amon in Karnak. It states that the king on ascending the throne found the temples of the country neglected and, as a result, the country forsaken by the gods and in confusion. The king therefore had to restore the temples. The description of this state of affairs may be a stereotype not necessarily having anything in particular to do with the preceding reign of Akhnaton, even if this historical fact may be alluded

88 Derchain 1965 p. 19; Assmann 1975 b p. 28.

to; this is not essential in our case. What is important is the way in which the text connects the neglected temples with the sad conditions of the land; it refers to a dependency of the gods upon the temple-cult for their presence in the land. We shall quote the relevant passage (in J. Bennett's translation)[89];

Now when His Majesty arose as king,
the temples of the gods and goddesses, beginning from
Elephantine [down] to the marshes of the Delta,
[their? --- had] fallen into neglect,
their shrines had fallen into desolation and become
tracts overgrown with K[3t?]3-plants,
their sanctuaries were as if they had never been,
their halls were a trodden path.
The land was in confusion, the gods forsook this land.
If an [army?] was sent to Djahy to widen the frontiers of
Egypt, it met with no success at all.
If one prayed to a god to ask things of him, [in no wise]
did he come.
If one made supplication to a goddess in like manner, in
no wise did she come.
Their hearts were weak of themselves (with anger);*
they destroyed what had been done[90].

The king then proceeds to enumerate what he has done for the temple-cult and declares that he has restored it so that the gods may protect Egypt *(T3-mrj)*. As a result of his activities

The gods and goddesses who are in this land, their hearts are joyful,
the possessors of shrines are glad,
lands are in a state of jubilation and merry-making,
exaltation is throughout [the whole land];
a goodly [state?] has come to pass[91].

* Bennett's parenthesis to *their hearts were weak (jb.sn fn)* is based on the assumption that the description is contrasted with *their hearts are in joy (jb.sn m ršwt)* (1.23), which appears likely. However, anger is not the only counter-part of joy, and in our case *weak hearts* might be interpreted as lack of vitality – in Egyptian thought the heart is the vital organ *per excellence.* (See Bonnet, *Reallexikon* p. 296 f.) Hearts *in joy* will, then, be indicative of joy of life.

89 Bennett 1939.

90 l. 5–10.

91 l. 23–24.

According to this text, the gods of Egypt are dependent upon the functioning of the temples. Without the temple-cult there are no gods present to uphold the cosmos of Egypt.

The theological frame of reference for this feature of the Egyptian concept of god is not immediately evident. But we can approach the cosmos-maintaining temple-cult in two ways, 1) from the iconographical point of view: we will consider the priest's role in the functioning of the temple, and 2) from the theological point of view: we will consider the cosmological concept of god formulated by this icon.

ad 1) In the temple, gods are created. We have seen in an earlier chapter that the priest acts the role of the theogon. This priest who creates gods is a phenomenon which can be explained from an iconographical point of view: when life is imparted to the gods of the temple, the priest actually belongs to the icon: that is, on the functional level this icon includes the priest. The ritual creation commences with the priest going into the temple; he literally enters the icon of the creator; he is assimilated with it in a similar way as the king is assimilated with his throne, the icon of Isis and becomes the son-of-the-goddess. The priest partakes in the creative nature of the temple; or, to put it differently, he assumes the role of the creator through being in the icon of the creator. It should be noted that the priest creates cosmos and the gods, only in the temple, and nowhere else. The divine quality of the icon is transferred to him. This is reflected in the words which he utters: he presents himself as a god[92]; and also in the acts he performs: he creates in ways which belong to the divine creator, namely by reciting sacred spells (cf. the *ḏ3jsw* of the cosmogony myth) and by transmitting the ka-life to the gods by embracing their statues – thus acting as the divine theogon[93]. In short, the temple lends to this representative of man the iconological quality of the divine creator. The priest creates gods through the temple, because this icon is the icon of the creator.

The Egyptians did not theologically expound the theogonic function of the priest, but there is an echo of it in the Hermetic literature, more precisely the *Asclepius,* in a passage of which treatise a reference is given to the priestly theogonic act which formulates it as a special phenomenon, even though the passage is coloured by the popular philosophy of its day (in A. J. Festugière's translation):

Et puisque voici annoncé le thème de la parenté et de la société qui lie hommes et dieux, connais donc, ô Asclépius, le pouvoir et la force de l'homme.

92 See p. 129 f.
93 See p. 120.

De même que le Seigneur et le Père ou, pour lui donner son nom le plus haut, Dieu, est le créateur des dieux du ciel, ainsi l'homme est-il l'auteur des dieux qui résident dans les temples et qui se satisfont du voisinage des humains: non seulement il reçoit la lumière (vie), mais il la donne à son tour, non seulement il progresse vers Dieu, mais encore il crée des dieux[94].

The similarity between man and the divine creator in this theogonic act is made explicit by the comparison with God, Father and Master. Trismegistus stresses the similarity by saying:

– comme le Père et Seigneur a doué les dieux d'éternité pour qu'ils lui fussent semblables, ainsi l'homme façonne-t-il ses propres dieux à la ressemblance de son visage.

To this statement the pupil, Asclepius, asks doubtfully: *Veux-tu dire les statues, ô Trismégiste?*
And Trismegistus drives home the answer in all its incredibility:

Oui, les statues, Asclépius. Vois comme toi-même tu manques de foi! Mais ce sont des statues pourvues d'une âme, conscientes, pleines de souffle vital, et qui accomplissent une infinité de merveilles; des statues qui conaissent l'avenir et le prédisent par les sorts, l'inspiration prophétique, les songes et bien d'autres méthodes, qui envoient aux hommes les maladies et qui les guérissent, qui donnent, selon nos mérites, la douleur et la joie[95].

Here is an attempt to explain the workings of the Egyptian cultic gods to the Hellenistic world. It operates with the model of an ideal and a real world of gods, but it will be observed that this model does not conform with that of a spiritual and a material world. The cultic gods made by man are material-and-spiritual, and they are living realities which affect the life of man. These are the gods with which man has his dealings. Trismegistus in no way tries to mitigate the bewildering impression that this concept of god apparently has on Asclepius, on the contrary, he stresses its peculiarity. In his answer to the hesitating Asclepius he shows himself as a "genuine Egyptian"[96].

In research, the explanations of the theogonic role of the priest have often applied some sort of magical theory. The term *magical* can be used with more than one meaning, and different aspects of the act termed *magical* can be seen as constitutive of its magical character. But if we choose a current definition,

94 Festugière and Nock 1960 p. 325.

95 *Op. cit.* p. 326.

96 As regards the close relationship between the *Asclepius* and Egyptian traditions, see Derchain 1962 a.

we might say that the magical act is an act which has the purpose of influencing a given course of events – through methods which are not scientifically grounded but rather associative in nature: employing principles of relations of determination based on similarity or on *pars pro toto*-representation, for instance mimetic gestures performed to effect the realization of that which they imitate, or uttering of names to effect the realization of their contents – the name belonging to that which it designates[97]. It is still a debated question whether the magical act should be considered a mechanical operation, i. e. an attempt to master the world (but without the necessary scientific insight to do so properly), for the magical act has an aspect of *opus operatum* which gives it the stamp of a technique for activating powers[98]. As magical acts are also found in religion, such as in cultic situations, some historians have been reluctant to regard magic as equivalent to science and may prefer to call the powers activated "*super*natural"[99]. The question whether the act relates to a supernatural or to a natural sphere is not, however, a question of choosing between a religious and a non-religious explanation, but a question of what kind of ontology is involved in the given case. Thus within a monistic ontology it does not make much sense to talk about "supernatural" powers.

In any case there is necessarily involved an ontology – a conception of nature is implied in the magical attempt to influence the events of life, and some historians have tried to explain the magical act from this ontological point of view. Ph. Derchain is among those who have contributed to the study of the Egyptian life-influencing rites from this angle[100].

97 These principles were first formulated by J. Frazer, and even if one may not agree with his general theory of magic, his principles may have a pragmatical value when applied to the formal structures of magical acts. We do not have to accept that the person who performs the acts has these principles in mind. A characterization of the formal structure of the magical acts should be included in their definition, or there will be no reason for using the term *magical* at all. A definition such as that of J. van Baal: "ritual acts, preferably of a simple character, executed to promote the realization of a concrete end" (*Symbols of Communication,* Assen 1971 p. 55) does not justify the use of the term; the difference between rites called magical and other rites would seem to lie in the impossible distinction between concrete and not concrete ends.

98 What I mean to say is that the act functions as if it were effective in itself. Whether the people performing it have doubted its effectiveness is another question which we have little means of answering – the psychology behind the act is in an extremely difficult area to explore.

99 Cf. Sharpe's definition of *magic* in his vocabulary for comparative religion: "Magic may be defined as the attempt on the part of an individual or group to influence a given course of events through the control or manipulation of supernatural forces", *50 Key Words,* Richmond 1971 p. 35.

100 Especially in *Le papyrus Salt 825.*

The magical acts found in cultic situations are those which attract our attention. In some religions, like the Egyptian, the magical element is indissolubly intertwined with the cultic act, or as H. Bonnet has appropriately expressed it: "So verwerben sich vollends Rel. und M.; Beschwörungen schlagen in Gebete um; Hymnen werden als Zaubersprüche gesprochen"[101]. It might not be incorrect to say that the typically Egyptian ritual act has this magical mark.

As regards these cultic, magical rites, one point should be made clear: as they are performed in a cultic situation they necessarily have theological implications. Therefore, it does not suffice to explain them from a purely ontological point of view. This is not to say that the ontological explanations are of no interest to the historian of religions, but if he gives an ontological explanation his task does not stop there. He also has to see the act in the light of its theological premises. Thus Ph. Derchain's attempt to explain the Egyptian rite by trying to trace a nature theory behind it is a useful effort but not sufficient for the historian of religions. Derchain describes the role of the priest and the functioning of the temple with reference to a dynamistic theory of nature: the temple is like "une centrale ou des énergies diverses sont converties en courant électrique, ou plus exactement à la salle des appareils de contrôle de cette centrale"[102]. He interprets the concept of god with reference to the same theory: "on peut dire que les dieux sont l'expression mythique des forces diffuses de l'univers, et leurs légendes celle des rapports de ces forces"[103].

The basic problem with this explanation is that it excludes the feature that these forces of nature are *gods* – an aspect of them which makes the issue automatically also a theological one. What Derchain says about the priest and the temple and the gods may prove sufficient from an ontological point of view taken by itself. But the historian of religions has to take into account that the priest, the temple, and the gods form a particular cultic constellation, which means that a theological meaning is inseparably included.

ad 2) There is a particular concept of god involved in the theogony performed by the priest: namely, that of the god who lives in man's world. Thus, when the priest – on entering the temple to create anew the life of the gods – informs them that "I am one of you" and calls himself by names of god, he

101 RÄRG p. 438; see also *op. cit.* p. 407.

102 *Le papyrus Salt 825,* p. 14. He explains the manipulation of the forces according to a theory of natural dynamism: "Ils seront ainsi de véritables transformateurs d'énergies, comme les hommes, mais infiniment plus puissants, capables de transformer l'énergie spirituelle en énergie matérielle" p. 12.

103 *Op. cit.* p. 12.

repeats in a cultic context the theological and cosmological fact that he moves in the world of gods.

On this ideological premise, an explanation of the cultic theogony can be formed which sees it in terms of a reciprocal divine-human relationship. It is an explanation which fits well into the cosmological ideology of the Edfu temple. As we have tried to demonstrate, the cosmos which is religiously relevant in the temple-context is not unspoilt nature but man-cultivated nature: man partakes in the creation of cosmos. There is therefore a fundamental inter-dependence between man and creator implied in the temple-ideology, as they both need the cosmos for their being. When this aspect of mutual dependence is focused upon in the ritual acts, the magical theory with its stress on man the manipulator is not adequate – the idea of divine-human interaction fades away. Unless we redefine the term *magical* to include the aspect of interaction, we might as well do without it.

The aspect of mutual dependence between man and god is especially evident in the rites which aim at creating and sustaining the life of the divine creator. The creator actually receives from the priest all that constitutes and maintains his life: the temple, the tracts of land supporting the temple, the statues, the insignia, the daily food and drink. All the prerequisites of his cultic life are offered to him in this place of offerings. But instead of the usual *do-ut-des*-explanation of these gifts[104], they might be explained with reference to the relationship of interdependence. *Do-ut-des* implies inter-dependence, but the theory formulates it with a unilaterial emphasis which leaves out the wider cult-ideological context in which giver and receiver appear to form a relationship of inseparable reciprocity. Within this context the entire temple-cult can be understood as man's contribution to this relationship – the one half of the enterprise, a ritual correlate to his cosmogonic activities which expresses the theological implications of these activities.

The notion of interaction is particularly prominent in the *maat*-offering. In the reliefs, this is depicted as an act by which the king-priest gives a little figure of the personified *maat* to the god[105]. *Maat* signifies the ordering principle of the world; it upholds the right order. Without *maat* there will be no life. Sometimes the word stands synonymous with life, and food-offerings can be interpreted as *maat*-offerings[106]. One might say that the offering of *maat* expresses what the whole temple-cult is about[107]: the maintenance of the world of gods and men. This is what the priest offers to the god:

104 E. g. E. Otto 1964 p. 84.
105 E. g. E I pl. XIIIb, XVIII, XX, XXIIb, XXVa, XXVIIa. Cf. Fairman 1958 p. 86–92.
106 Moret 1902 pp. 138–165; Bergman 1968 p. 216 f.; Assmann 1970 p. 63 f.
107 Hornung 1973 p. 211.

the capacity to uphold the world[108]. The offering supports the communal life[109], and it presupposes reciprocity rather than *do-ut-des,* because the king already has received the life-sustaining capacity – this is precisely what he shows by offering it to the god who is its source[110]. In this interpretation we are in agreement with E. Hornung who has described the *maat*-offering in the following way: "Was bei der Schöpfung von den Göttern kam, die Maat, kehrt aus der Hand des Menschen zu ihnen zurück – Symbol für die Partnerschaft zwischen Gott und Mensch, wie sie die ägyptische Religion verwirklicht hat. Aus dieser Partnerschaft, aus gegenseitiger Wirkung und Antwort, erklärt sich des sonst unlösbare Ineinander von Willensfreiheit und Vorherbestimmung. Durch die Schöpfung ist Göttern und Menschen eine gemeinsame Aufgabe gesetzt: ... zusammen an einer lebendigen Ordnung zu bauen, die dem Schöpferatem Raum gibt und nicht der Erstarrung verfällt"[111].

Another expression of the temple as a place of offering the life to the creator is the nome-processions walking towards the god of the temple with their food. They are headed by the king; he is the one that leads them to the god[112], giving that which sustains the life to the creator: the produce of the man-cultivated cosmos.

In this perspective of human-divine reciprocity, the priest plays the role of the *king,* the representative of man as society. As king he is instituted to be responsible for man's part in the creation of cosmos[113]. Also, he is responsible for man's part in building the temple – the cultic correlate to cosmos. In the latter function he is – as in the first – equal with the gods; as we have seen, the king is depicted in the reliefs to the foundation ritual working together with the creator gods.

Thus the cosmological identity of the creator can indicate a way of explaining the modes in which divine life is represented in the temple and created by its rituals. God represents a cosmos which is shaped by man: canalized water, irrigated land, mounds upon which the houses are built. The temple is the icon of this god who comes into being when man takes part in the cosmogony; the cosmic "body" of the creator, in which gods and men have their being, presupposes the creative contribution of man; it is also man's achievement.

108 *Ad maat* as a cosmological principle, see Assmann 1969 p. 267 f.
109 See E. Otto 1964 p. 63 ff.; Derchain 1962 c and 1965 p. 14.
110 See E. Otto 1964 p. 26; Assmann 1969 p. 157.
111 Hornung 1973 p. 211 f.
112 Beinlich 1972 pp. 30, 33.
113 Assmann 1970 p. 58 f.

Still, what can *not* be explained from this point of view is the fact that the temple and its cult are actually thought to influence the cosmos: that the cultic life of god is considered to be an objective reality. In other words, the specific iconological function of realizing the presence of god in a way that affects the cosmos it represents is left unaccounted for. The temple is more than an expression of the divine cosmos and its workings; it is also an icon, the corporeal god of the cosmos, and this cultic phenomenon has actual cosmogonic power. When Tutankhamon states that the temples of Egypt lay in ruins, he at the same time states that the Egyptian cosmos has dissolved; it is laid open for the enemy, because the cosmos-sustaining gods have turned their backs on it – a consequence of the ruined temples and lack of temple-cult. The thought is also reflected in the Kom Ombo-hymn recited when the priest enters the temple to create the cosmic life of the gods, saying that insofar as the temple exists, the earth and the horizon will exist, and *vice versa*[114]. To explain this functional feature of the icon would, however, require a wider range of vision than this monograph applies, namely one that includes the entire ontology of the Egyptian culture.

It has been demonstrated that the creator and his gods exist in a symbiosis with a particular society and a particular culture. This fact, expressed also on the Tutankhamon-stela, long occupied the minds of Egyptians; it was even echoed by those who thought about their religion within the framework of Hellenistic idealism and transcendentalism – as can be seen in the Hermetic literature. Thus it is said in the *Asclepius* that when the Egyptian culture has gone, the Egyptian gods will have gone, too. We shall quote the relevant passage (in Festugière's translation). The words are put in the mouth of Trismegistus:

Ignores-tu donc, Asclépius, que l'Egypte est la copie de ciel ou, pour mieux dire, le lieu où se transfèrent et se projettent ici-bas toutes les opérations que gouvernent et mettent en oeuvre les forces célestes? Bien plus, s'il faut dire tout le vrai, notre terre est le temple du monde entier.

Et cependant, puisqu'il convient aux sages de connaître à l'avance toutes les choses futures, il en est une qu'il faut que vous sachiez. Un temps viendra où il semblera que les Egyptiens ont en vain honoré leurs dieux, dans la piété de leur coeur, par un culte assidu: toute leur sainte adoration échouera inefficace, sera privée de son fruit. Les dieux, quittant la terre, regagneront le ciel; ils abandonneront l'Egypte; cette contrée qui fut jadis le domicile des saintes liturgies, maintenant veuve de ses dieux, ne jouira plus de leur présence. Des étrangers rempliront ce pays, cette terre, et non seulement on n'aura plus soici des

114 See p. 129.

observances, mais, chose plus pénible, il sera statué par de prétendues lois, sous peine de châtiments prescrits, de s'abstenir de toute pratique religieuse, de tout acte de piété ou de culte envers les dieux. Alors cette terre très sainte, patrie des sanctuaires et des temples, sera toute couverte de sépulcres et de morts. O Egypte, Egypte, il ne restera de tes cultes que des fables et tes enfants, plus tard, n'y croiront même pas; rien ne survivra que des mots gravés sur les pierres qui racontent tes pieux exploits. Le Scythe, ou l'Indien, ou quelque autre pareil, je veux dire un voisin barbare, s'établira en Egypte. Car voici que la divinité remonte au ciel; les hommes, abandonnés, mourront tous, et alors, sans dieu et sans homme, l'Egypte ne sera plus qu'un désert[115].

This presentation of the Egyptian gods and their relations to Egypt is interesting in more than one respect. Even though it belongs to a Hellenistic milieu, it has an unmistakably Egyptian stamp. First of all, we can note the idea that the gods live in Egypt[116] *le temple du monde entier;* only here the idea is not any longer taken for granted, as in the Edfu material, and Trismegistus has obvious problems in finding an adequate formulation – he tries several, dealing with the problem of making himself understood, having realized that in this theological feature the Egyptian religion might differ from others of his day.

Our attention is also drawn to the fact that the passage documents the life of the gods as intertwined with man's activities, so closely that the Egyptian gods and the Egyptian way of life presuppose each other: when the one has gone, the other has gone, too. Whether it is this theology of cosmos which the author sees on its vane, or the unity of god and culture, is another question which we will not venture to discuss here.

To prevent a reductionistic conclusion from this discussion on the intimate relationship between the gods and this particular cosmos which the temple witnesses, I will repeat what I have said above: the temple is an icon; it presents god as actual presence, a cultic phenomenon of objective influence. Being an icon the temple is more than a description of the Egyptian world; it is a phenomenon *sui generis.* The monograph has refrained from giving an ontological explanation of the phenomenon. But we have arrived at an iconological characterization of the temple as the dynamic, cultic presence of the all-comprehensive creator of the Egyptian world.

It remains to be said that this icon is a typically Egyptian creation. That it dates from a period which introduces a diametrically different concept of god, is one of history's many ironies.

115 Festugière and Nock 1960 p. 326 f.

116 The idea was given its own history in the Hellenistic tradition about Egypt as the paradise of gods, see Tardieu 1974 p. 274.

Works cited

Alliot, M.

1944 *Le culte d'Horus à Edfou au temps des Ptolémées,* BdE XX fasc. II, Cairo.

1949 *Le culte d'Horus à Edfou au temps des Ptolémées,* BdE XX fasc. I, Cairo.

1966 ed. A. Barucq, „Les textes cosmogoniques d'Edfou d'après des manuscrits laissés par Maurice Alliot, présentés par André Barucq", BIFAO 64.

Arnold, D.

1962 *Wandrelief und Raumfunktion in ägyptischen Tempeln des Neuen Reiches,* MÄS 2.

Assmann, J.

1969 *Liturgische Lieder an den Sonnengott. Untersuchungen zur altägyptischen Hymnik,* MÄS 19.

1970 *Der König als Sonnenpriester,* ADAIK 7.

1975a *Ägyptische Hymnen und Gebete,* Munich.

1975b „Zeit und Ewigkeit im alten Ägypten", AHAW Phil.-hist. Kl.

1979 „Primat und Transzendenz", in *Aspekte der spätägyptischen Religion,* ed. W. Westendorf (= Göttinger Orientforschung IV, Ägypten B. 9) Wiesbaden.

1980 „Grundstrukturen der ägyptischen Gottesvorstellungen", *Biblische Notizen* II.

Aufrère, S.

1982–1983 „Caractères principaux et origine des minéraux", RdE 34.

Badawy, A.

1968 *A History of Egyptian Architecture* III, Berkeley and Los Angeles.

Baldwin Smith, E.

1968 *Egyptian Architecture as Cultural Expression,* New York.

Barguet, P.

1980 „La cour du temple d'Edfou et le cosmos", IFAO: *Livre du centenaire 1880–1980.*

Beinlich, H.

1976 *Studien zu den geographischen Inschriften,* Bonn.

Bennett, J.

1939 „The Restoration Inscription of Tutankhamon", JEA 25.

Bergman, J.

1968 *Ich bin Isis,* Uppsala.

1970 „*B3* som gudomlig uppenbarelsesform i det gamla Egypten", in *Religion och Bibel* 29, Uppsala.

Blackman, A. M. and H. W. Fairman
1935–1944 „The Myth of Horus at Edfu", JEA 21 (1935), 28 (1942), 29 (1943), 30 (1944).
1941 „A Group of Texts inscribed on the Façade of the Sanctuary in the temple of Edfu", *Miscellanea Gregoriana,* Rome.
1946 „The Consecration of an Egyptian Temple According to the Use of Edfu", JEA 32.

Bleeker, C. J.
1973 *Hathor and Thoth,* Leiden.

Bonneau, D.
1964 *La crue du Nil,* Paris.

Borchardt, L.
1897 *Die ägyptische Pflanzensäule,* Berlin.
1902–1903 „Die Cyperussäule", ZÄS 40.

Boylan, P.
1922 *Thoth, the Hermes of Egypt,* London.

Breasted, J. H.
1912 *Development of Religion and Thought in Ancient Egypt,* New York.

Brugsch, H.
1879 *Dictionnaire Géographique de l'Ancienne Egypt,* Leipzig.

Brunner, H.
1955 a „Die Grenzen von Zeit und Raum bei den Ägyptern", AfO 17.
1955 b „Zum Zeitbegriff der Ägypter", StG 8.
1957 „Zum Raumbegriff der Ägypter", StG 10.
1970 „Die Sonnenbahn in ägyptischen Tempeln", in *FS Kurt Galling,* Tübingen.

de Buck, A.
1922 *De Egyptische Voorstellingen betreffende den Oerheuvel,* Leiden.
1948 „On the meaning of the name *Ḥcpj*", ON.

Butzer, K. W.
1978 „Perspectives on Irrigation Civilizations", in *Immortal Egypt,* ed. D. Schmandt-Besserat, Malibu.

Capart, J.
1904 „Sur le prêtre *'Iwn-mwt.f*", ZÄS 41.

Chassinat, E.
1929–1934 *Le Temple d'Edfou* IV–XIV, MMAF 21–31.
1934–1965 *Le Temple de Dendara* I–VI, Cairo.
1939 *Mammisi d'Edfou,* MIFAO 16.
1966 *Le mystère d'Osiris au mois de Khoiak* I, Cairo.

Chassinat, E. and M. de Rochemonteix
1918, 1928 *Le Temple d'Edfou* II–III, MMAF 11, 20.

Daumas, F.
1951 „Sur trois représentations de Nout à Dendara", ASAE 51.

1956 „La valeur de l'or dans la pensée religieuse égyptienne", RHR 149.

1981 „L'interprétation des temples égyptiens anciens à la lumière des temples gréco-romains", *Karnak* VI.

Derchain, Ph.

1962 a „L'authenticité de l'inspiration égyptienne dans le ‚Corpus Hermeticum'", RHR.

1962 b „Un manuel de géographie liturgique à Edfou", CdE 37 No. 73.

1962 c „Le rôle du roi d'Egypte dans le maintien de l'ordre cosmique", in *Le pouvoir et le sacré* (Centre d'étude des religions I) Brussels.

1965 *La papyrus Salt 825, rituel pour la conservation de la vie en Egypte,* Mémoire de l'académie Royale de la Belgique, Classe de Lettres 58, fasc. 1 a) Brussels.

1966 „Réflexions sur la décoration de pylones", BSFE 46.

1978 „En l'an 363 de sa Majesté le Roi...", CdE 53.

Dondelinger, E.

1973 *Der Jenseitsweg der Nofretari,* Graz.

Eliade, M.

1959 *Cosmos and History. The Myth of the Eternal Return,* New York.

1964 *Myth and Reality,* London.

Erman, A.

1934 *Die Religion der Ägypter,* Berlin and Leipzig.

Fairman, H. W.

1943 „Notes on the Alphabetic Signs employed in the Hieroglyphic Inscriptions of the Temple of Edfu", ASAE 43.

1944 „Ptolemaic Notes", ASAE 44.

1945 „An Introduction to the Study of Ptolemaic Signs and their Values", BIFAO 43.

1954 „Worship and Festivals in an Egyptian Temple", Bulletin of the John Ryland's Library 37, Manchester.

1958 „A Scene of Offering the Truth in the Temple of Edfu", MDAIK 16.

1960 „The Kingship Rituals of Egypt", in *Myth, Ritual, and Kingship,* ed. S. H. Hooke, Oxford.

Faulkner, R. O.

1938 „The pap. Bremner-Rhind", JEA 24.

1972 *The Book of the Dead,* New York.

Festugière, A.-J. and A. D. Nock

1960 *Corpus Hermeticum* II, *Asclepius,* Paris.

Forbes, R. J.

1955 „Irrigation of ancient Egypt", in *Studies in ancient technology* II, Leiden.

Frankfort, H.
1948 *Ancient Egyptian Religion,* New York.
1969[6] *Kingship and the Gods,* Chicago and London (1948).
Gallery, L. M.
1978 „The Garden of Ancient Egypt", in *Immortal Egypt,* ed. D. Schmandt-Besserat, Malibu.
Gardiner, A.
1906 „Mesore as first month of the Egyptian year", ZÄS 43.
1950 „The Baptism of Pharaoh", JEA 36.
Gauthier, H.
1925–1931 *Dictionnaire des Nomes Géographiques,* 1–7, Cairo.
Gothein, M. L.
1966[2] *A History of Garden Art* vol. I, ed. W. P. Wright, New York.
Grapow, H.
1931 „Die Welt vor der Schöpfung", ZÄS 67.
Griffiths, J. G.
1960 *The Conflict of Horus and Seth,* Liverpool.
1970 (ed.), (Plutarch), *De Iside et Osiride,* Cambridge.
Gutbub, A.
1953 „Jeux de signes dans quelques inscriptions des grands temples de Dendérah et d'Edfou", BIFAO 52.
1961 „Hathor *ḫnt 'Iwn.t,* Re Hor *ḫnt Bḥdt,* Amon *ḫnt W3s.t*", *Mélanges Mariette,* BdE 32.
1973 *Textes fondamentaux de la théologie de Kom Ombo,* BdE 47.
Helck, W.
1976 „Die Systematik der Ausschmückung der hypostylen Halle von Karnak", MDAIK 32.
Holmberg, M. Sandman
1946 *The God Ptah,* Lund.
Hooke, S. H. (ed.)
1960 *Myth, Ritual, and Kingship,* Oxford.
Hornung, E.
1956 „Chaotische Bereiche in der geordneten Welt", ZÄS 81.
1966 *Geschichte als Fest,* Darmstadt.
1963, 1967 *Das Amduat. Die Schrift des verborgenen Raumes* I, II: Ägyptologische Abhandlungen 7, Wiesbaden 1963, III: ÄA 13, 1967.
1972 Ägyptische Unterweltsbücher (Die Bibliothek der alten Welt), München and Zürich.
1973 *Der Eine und die Vielen,* Darmstadt.
1974 „Seth. Geschichte und Bedeutung eines ägyptischen Gottes", *Symbolon* 2, Köln.
1976 *Das Buch der Anbetung des Re im Westen* (= Ägyptiaca Helvetica 3).

1979 a *Das Totenbuch der Ägypter* (Die Bibliothek der alten Welt), München and Zürich.

1979 b „Die Tragweite der Bilder Altägyptische Bildaussagen", *Eranos: Denken und Mytische Bildwelt.*

1981 „Zu den Schluss-szenen der Unterweltsbücher", *Festschrift für Labib Habachi,* MDAIK 37.

Jequier, G.

1924 *Les temples ptolémaïques et romains,* Paris.

1946 *Considération sur les religions égyptiennes,* Neuchatel.

Junge, F.

1978 „Wirklichkeit und Abbild", in *Synkretismusforschung – Theorie und Praxis,* ed. G. Wiessner, Wiesbaden.

Junker, H.

1910 „Schlacht- und Brandopfer und ihre Symbolik im Tempelkult der Spätzeit", ZÄS 47.

1912 „Der Bericht Strabos über den heiligen Falken von Philae im Lichte der ägyptischen Quellen", WZKM 26.

1913 „Das Götterdekret über das Abaton", DAWW 41.

Kahler, E.

1966 „The Nature of the Symbol", in *Symbolism in Religion and Literature,* ed. R. May, New York.

Kees, H.

1943 „Farbensymbolik in ägyptischen religiösen Texten", NAWG Phil. hist. Kl., Göttingen.

1956[2] *Totenglauben und Jenseitsvorstellungen der alten Ägypter,* Leipzig (1926).

1977[3] *Der Götterglaube im Alten Ägypten,* MVÄG 45 (1941).

Kuentz, Ch.

1928–1934 *La bataille de Quadesh,* MIFAO 55.

Kuhlmann, K. P.

1977 *Der Thron im alten Ägypten,* ADAIK 10.

Kurth, D.

1975 *Den Himmel stützen* (Rites égyptiens II), Brussels.

1983 *Die Dekorationen der Säulen im Pronaos des Tempels von Edfu* (= Göttinger Orientforschungen, Reihe IV: Ägypten B. 11) Wiesbaden.

Labrique, F.

1982 „Observations sur le temple d'Edfou", GM 58.

Lange, K. and M. Hirmer

1955 *Ægypten,* München.

Lipinski, E.

1979 *State and temple economy in the Ancient Near East* I, OLA 5.

Luft, U.
1978 *Beiträge zur Historisierung der Götterwelt und der Mythenschreibung,* Stud Aeg IV Budapest.

Mariette, A.
1870–1880 *Dendérah. Description Générale du Grand Temple de cette ville* I–V, Paris.

Meeks, D.
1972 *Le grand texte des donations au temple d'Edfou,* BIFAO 59.

Mohiy el-Din Ibrahim
1979 „The God of the Great Temple of Edfou", in *Studies in Honour of H. W. Fairman,* Warminster.

Montet, P.
1957–1961 *Géographie de l'Egypte ancienne* I: Paris 1957, II: Paris 1961.
1964 „Le rituel de fondation des temples égyptiens", Kêmi 17.

Morenz, S.
1960 *Ägyptische Religion* (= Die Religionen der Menschheit 8) Stuttgart.
1975 „Wortspiele in Ägypten", in *Religion und Geschichte des alten Ägypten,* eds. E. Blumenthal and S. Herrmann, Köln and Wien.

Moret, A.
1902 *Le rituel du culte divin journalier en Egypte,* Paris.

Müller, D.
1967 „Neue Urkunden zur Verwaltung im Mittleren Reich", *Orientalia* 36.

Nelson, H. H.
1942 „The identity of Amon-Re of United-with-Eternity", JNES I.

Noshy, I.
1937 *The Arts in Ptolemaic Egypt,* London.

Otto, E.
1938 „Die beiden Länder Ägyptens in der ägyptischen Religionsgeschichte", Stud Aeg I, Rome (= Analecta orientalia 17).
1958 *Das Verhältnis von Rite und Mythos im Ägyptischen,* Heidelberg.
1960 *Das ägyptische Mundöffnungsritual* (= Ägyptologische Abhandlungen 3), Wiesbaden.
1964 *Gott und Mensch nach den ägyptischen Tempelinschriften der gr.-röm. Zeit,* Heidelberg.
1966 „Zeitvorstellungen und Zeitrechnung im Alten Orient", StG 19.
1969 *Wesen und Wandel der ägyptischen Kultur,* Berlin, Heidelberg, New York.

Otto, W.
1905 *Priester und Tempel im hellenistischen Aegypten,* Leipzig und Berlin.

Parker, R. A.
1950 *The Calendars of Ancient Egypt* (= Studies in Ancient Oriental Civilization 26), Chicago.

Piankoff, A.
1955 *The Shrines of Tut-ankh-amon* (Egyptian religious texts and representations 2), New York.

Porter, B. and R. L. B. Moss
1939 *Topographical Bibliography of Ancient Egyptian Hieroglyphic Texts, Reliefs and Paintings* VI, Oxford.

Reymond, E. A. E.
1962 a „The Primeval Djeba", JEA 48.
1962 b (E. A. E. Jelinkova), „The Shebtiw in the temple of Edfou", ZÄS 87.
1963 „Worship of the Ancestor Gods at Edfu", CdE 75.
1963–1964 „The Origin of the Spear" I: JEA 49 (1963), II: JEA 50 (1964).
1965 „The Cult of the Spear in the Temple at Edfu", JEA 51.
1967 „The God's *ỉḫt*-relics", JEA 53.
1966, 1969 „The Children of Tanen" I: ZÄS 92 (1966), II: ZÄS 96 (1969).
1969 *The Mythical Origin of the Egyptian Temple*, Cambridge.

Ricke, H.
1944 *Beiträge zur ägyptischen Bauforschung und Altertumskunde* 4, Zürich.

Rochemonteix, M. de
1894 „Le temple égyptien", Bibliothèque Egyptologique 3.
1897 *Le temple d'Edfou* I, MMAF 10.

Rochemonteix, M. de and E. Chassinat
1918, 1928 *Le temple d'Edfou* II, II, MMAF 11, 20.

Roeder, G.
1959 *Die ägyptische Götterwelt* (Die Bibliothek der alten Welt), Stuttgart and Zürich.
1960 a *Kulte, Orakel und Naturverehrung in den alten Ägypten* (Die Bibliothek der alten Welt), Stuttgart and Zürich.
1960 b *Mythen und Legenden um ägyptischen Gottheiten und Pharaonen* (Die Bibliothek der alten Welt), Stuttgart and Zürich.

de Rougé, J.
1865–1874 „Textes géographiques du temple d'Edfou", RAr 11–28.

Säve-Söderbergh, T.
1953 „On Egyptian Representations of Hippopotamus Hunting as a Religious Motif", *Horae Soederblomianae* III, Uppsala.

Sauneron, S.
1957 „A propos de deux signes ‚ptolémaiques'", BIFAO 56.
1958 „L'abaton de la campagne d'Esna", MDAIK 16.
1959–1982 *Esna* I–VIII, PIFAO.

1966 „Le monde du magicien égyptien", in *Le monde sorcier* (= Sources orientales VII), Paris.

Sauneron, S. and H. Stierlin
1978 *Die letzten Tempel Ägyptens; Edfu und Philae,* Zürich.

Sauneron, S. and J. Yoyotte
1959 „La naissance du monde selon d'Egypte ancienne", in *La naissance du monde* (= Sources orientales I), Paris.

Schäfer, H.
1943 „Die ‚Vereinigung der beiden Länder'", MDAIK 12.
1957 „Das Niederschlagen der Feinde", WZKM 54.

Schenkel, W.
1977 *Kultmythos und Märtyrerlegende,* Göttinger Orientforschung IV. Reihe: Ägypten B. 5, Wiesbaden.

Schlott, A.
1969 *Die Ausmaße Ägyptens nach altägyptischen Texten,* Darmstadt.

Schnebel, M.
1925 *Die Landwirtschaft im hellenistischen Ägypten* I (= Münchner Beiträge zur Payrusforschung und antiken Rechtsgeschichte 7), München.

Schott, S.
1950 „Altägyptische Festdaten", AAWLM 10.
1963 „Die Opferliste als Schrift des Thot", ZÄS 90.

Schweitzer, U.
1956 *Das Wesen des Ka* (= Ägyptologische Forschungen 19), Glückstadt, Hamburg, New York.

Sethe, K.
1908–1922 *Die altägyptischen Pyramidentexte,* Leipzig.
1929 „Amun und die acht Urgötter von Hermopolis", APAW.

Steindorff, G.
1896 „Haus und Tempel, ZÄS 34.

Täckholm, v. and M. Drar
1950 *The Flora of Egypt* II (= Cairo University Bulletin of the Faculty of Science 28), Cairo.

Tardieu, M.
1974 *Trois mythes gnostiques,* Paris.

Traunecker, C.
1972 „Les rites de l'eau à Karnak d'après les textes de la rampe de Taharqa", BIFAO 72.

Watterson, B.
1979 „The Use of Alliteration in Ptolemaic", *Studies in Honour of H. W. Fairman,* Warminster.

Wiedemann, A.
1920 *Das alte Ägypten* (Kulturgeschichtliche Bibliothek 2), Heidelberg.

Wilke, C.
1934 „Zur Personifikationen von Pyramiden", ZÄS 70.

Winter, E.
1968 *Untersuchungen zu den ägyptischen Tempelreliefs der griechisch-römischen Zeit,* ÖAW Denkschriften 98, Wien.
1969 „Zwei Beobachtungen zur Formung des ägyptischen Tempelreliefs in der griechisch-römischen Zeit", *Religions en Egypte hellénistique et romaine,* Paris.

de Wit, C.
1956 „Some Values of Ptolemaic Signs", BIFAO 55.
1961 „Inscription dédicatoires du temple d'Edfou", CdE 71, 72.

Žabkar, L. V.
1968 *A Study of the Ba-conception in Ancient Egyptian Texts,* Chicago.
1981 „Hymn to Osiris Pantocrator", ZÄS 108.

Text

The cosmogony text on the inner face of the enclosure wall.
From Chassinat, *Le temple d'Edfou* VI 181,5–186,10.

TABLEAU I'n. 3 d. II (pl. CXLIX).

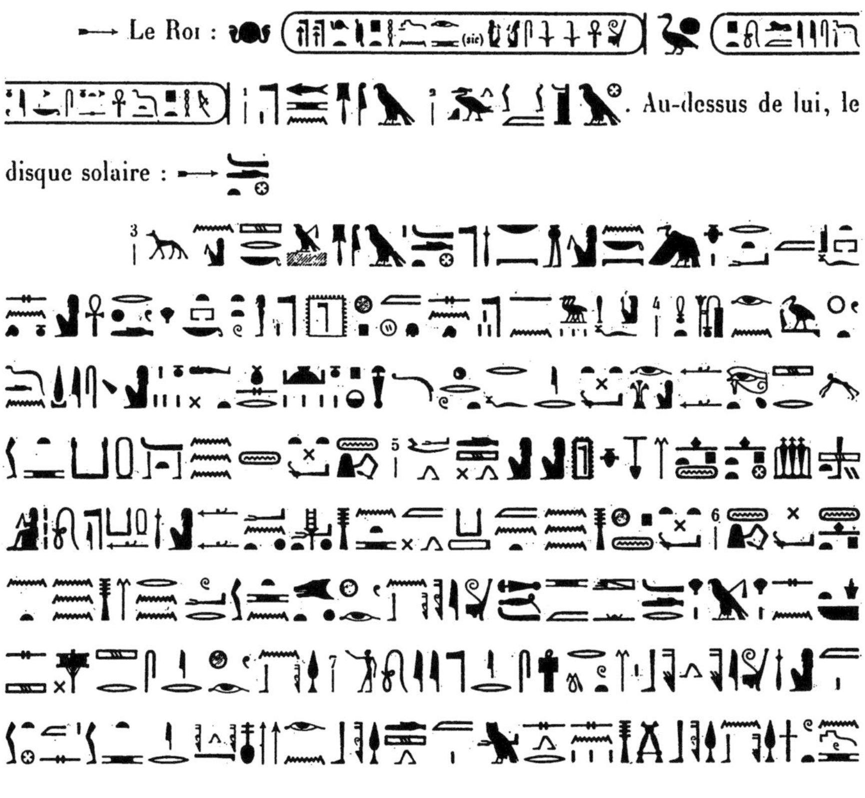

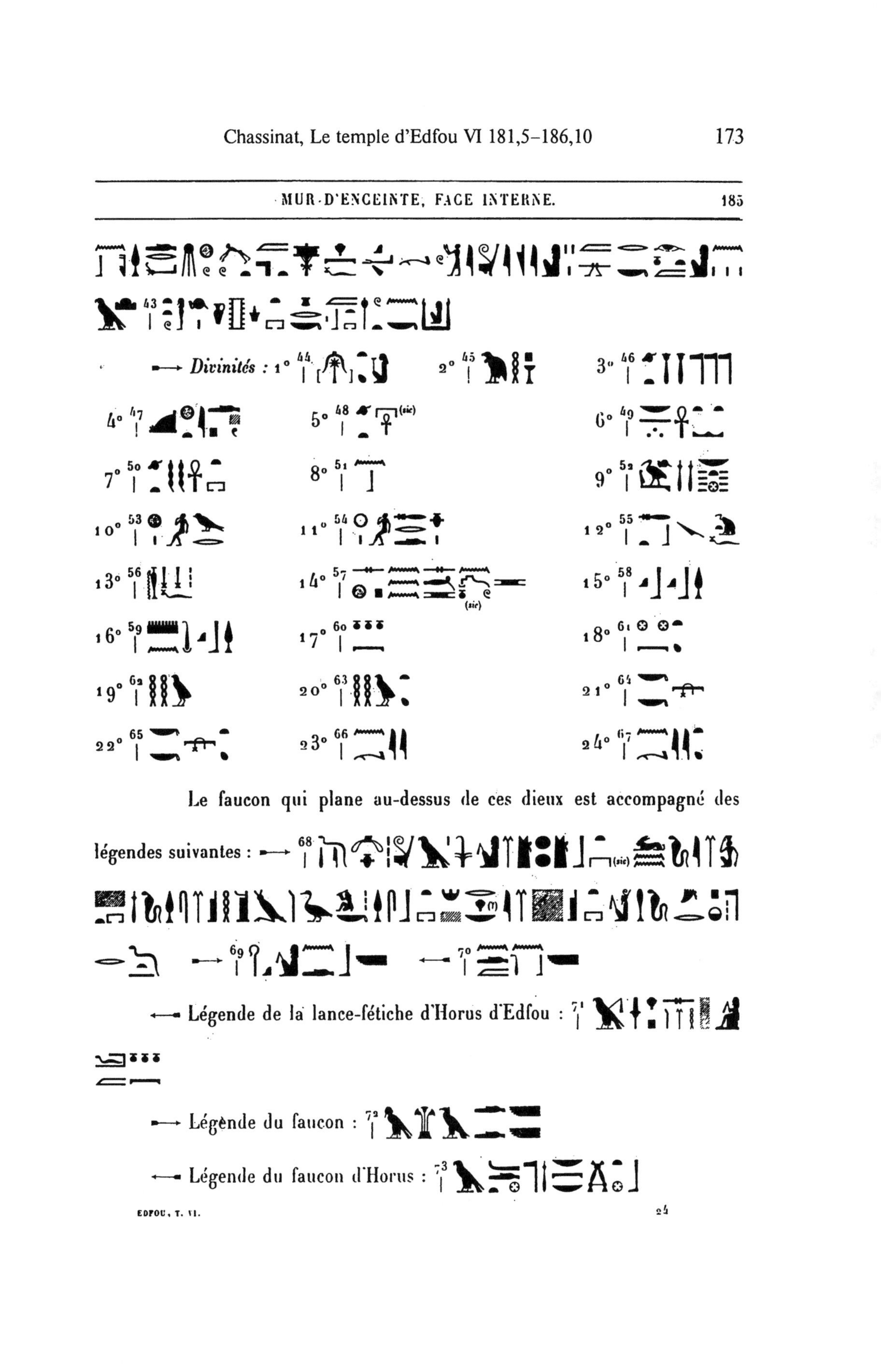

43

→ *Divinités* : 1° 44 — 2° 45 — 3° 46 — 4° 47 — 5° 48 (sic) — 6° 49 — 7° 50 — 8° 51 — 9° 52 — 10° 53 — 11° 54 — 12° 55 — 13° 56 — 14° 57 (sic) — 15° 58 — 16° 59 — 17° 60 — 18° 61 — 19° 62 — 20° 63 — 21° 64 — 22° 65 — 23° 66 — 24° 67

Le faucon qui plane au-dessus de ces dieux est accompagné des légendes suivantes : → 68 (sic)

→ 69 ← 70

← Légende de la lance-fétiche d'Horus d'Edfou : 71

→ Légende du faucon : 72

← Légende du faucon d'Horus : 73

Génies protecteurs d'Edfou : 1° ⟶ 

2° ⟵ 

⟵ Le dieu du temple d'Edfou : 

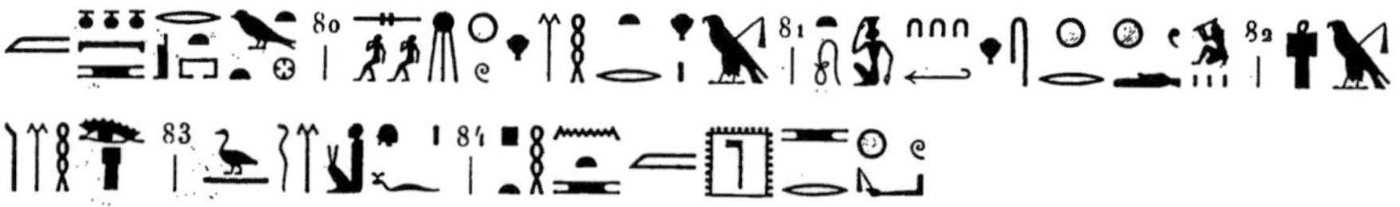

⟵ Pe-Khen : 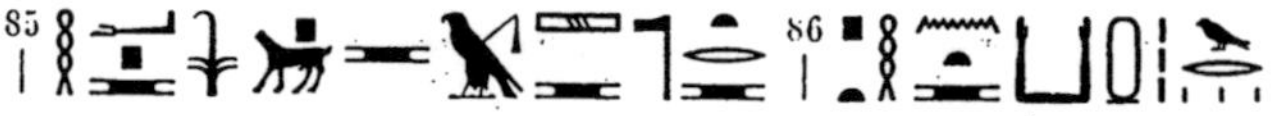

TABLEAU I'n. 3 d. III[3] (pl. CXLIX).

⟵ Le Roi : 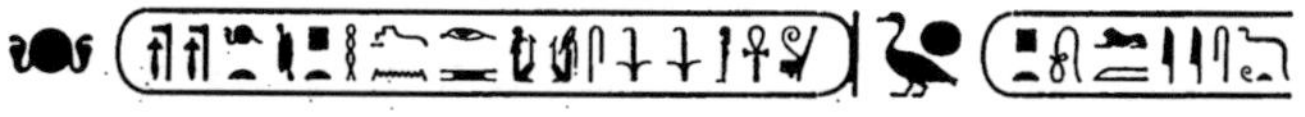

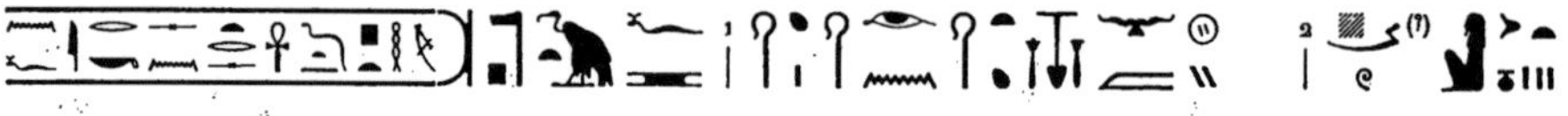

. Au-dessus de lui, le disque solaire : ⟵ 

⟶ Horus fils d'Isis : 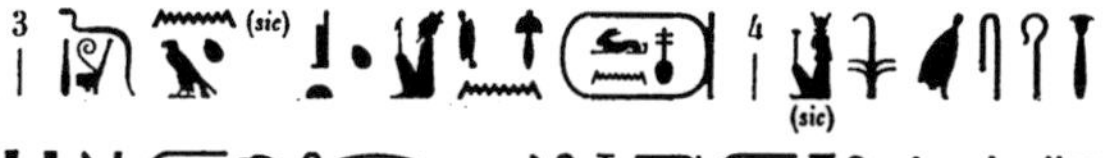

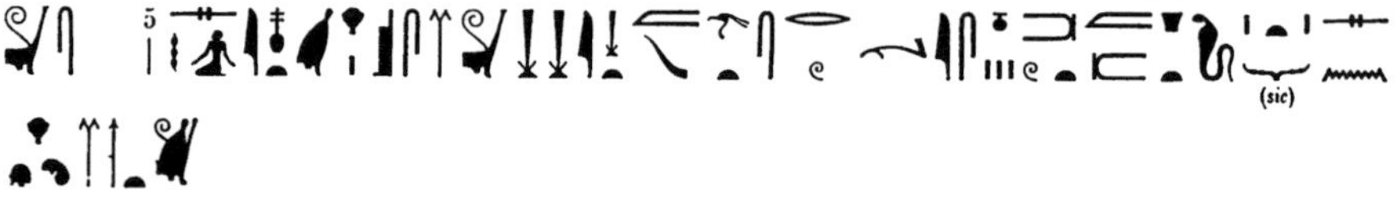

[1] La ligne 91 a été publiée par Lepsius, *Denkm.*, Abth. IV, pl. 46 *b*.

[2] Lepsius : ›, pour ❩.

[3] Les lignes 9-28 de ce tableau ont été publiées par Lepsius, *Denkm.*, Abth. IV, pl. 46 *b*.

Illustrations

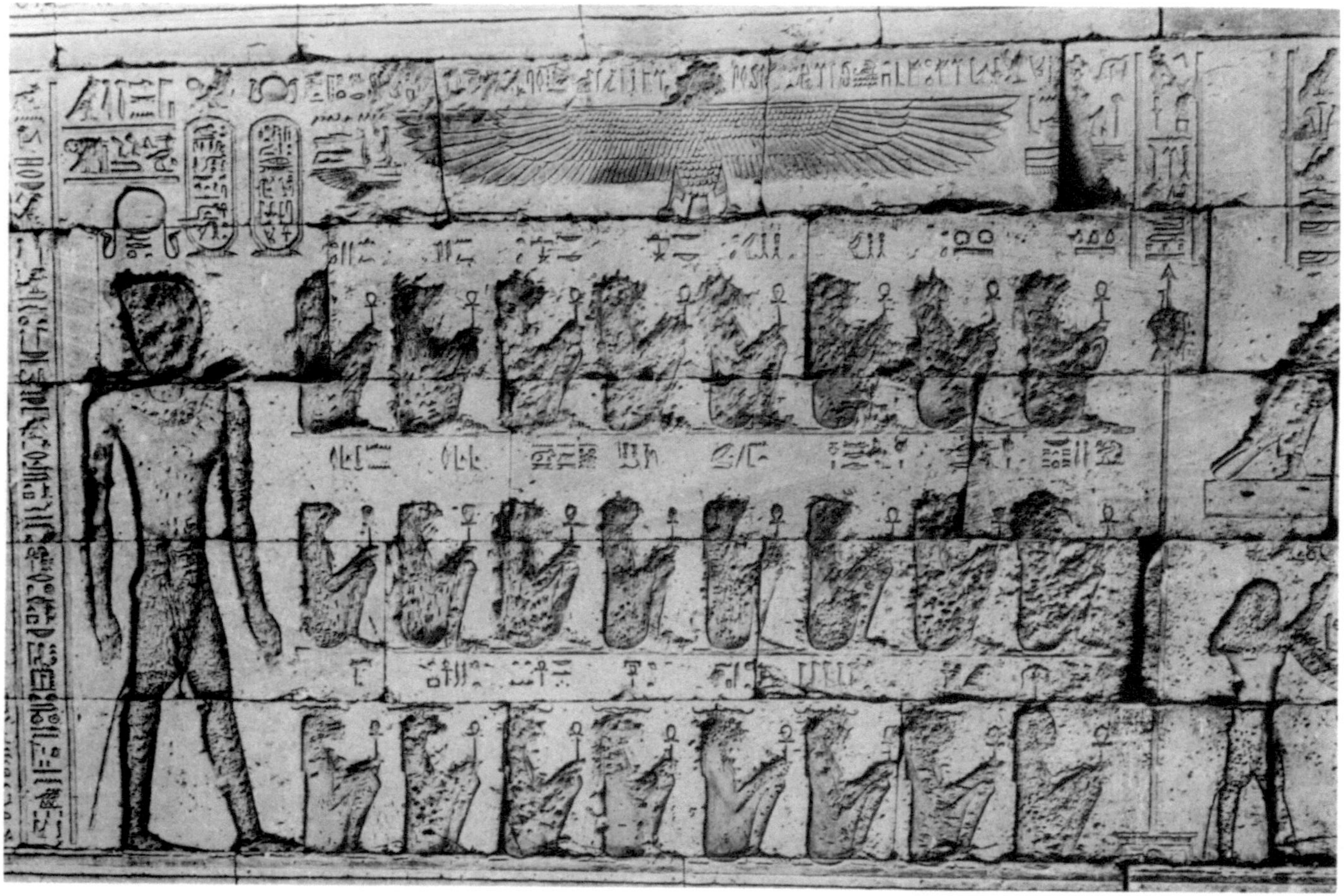

Plate 1: Reliefs to the cosmogony text on the inner face of the enclosure wall. Assembly of gods facing the cultic weapon of Horus. *Above,* the Lord-of-flying spreads out his wings. *Far left,* the king.
From Chassinat, *Le temple d'Edfou* XIV pl. dlx.

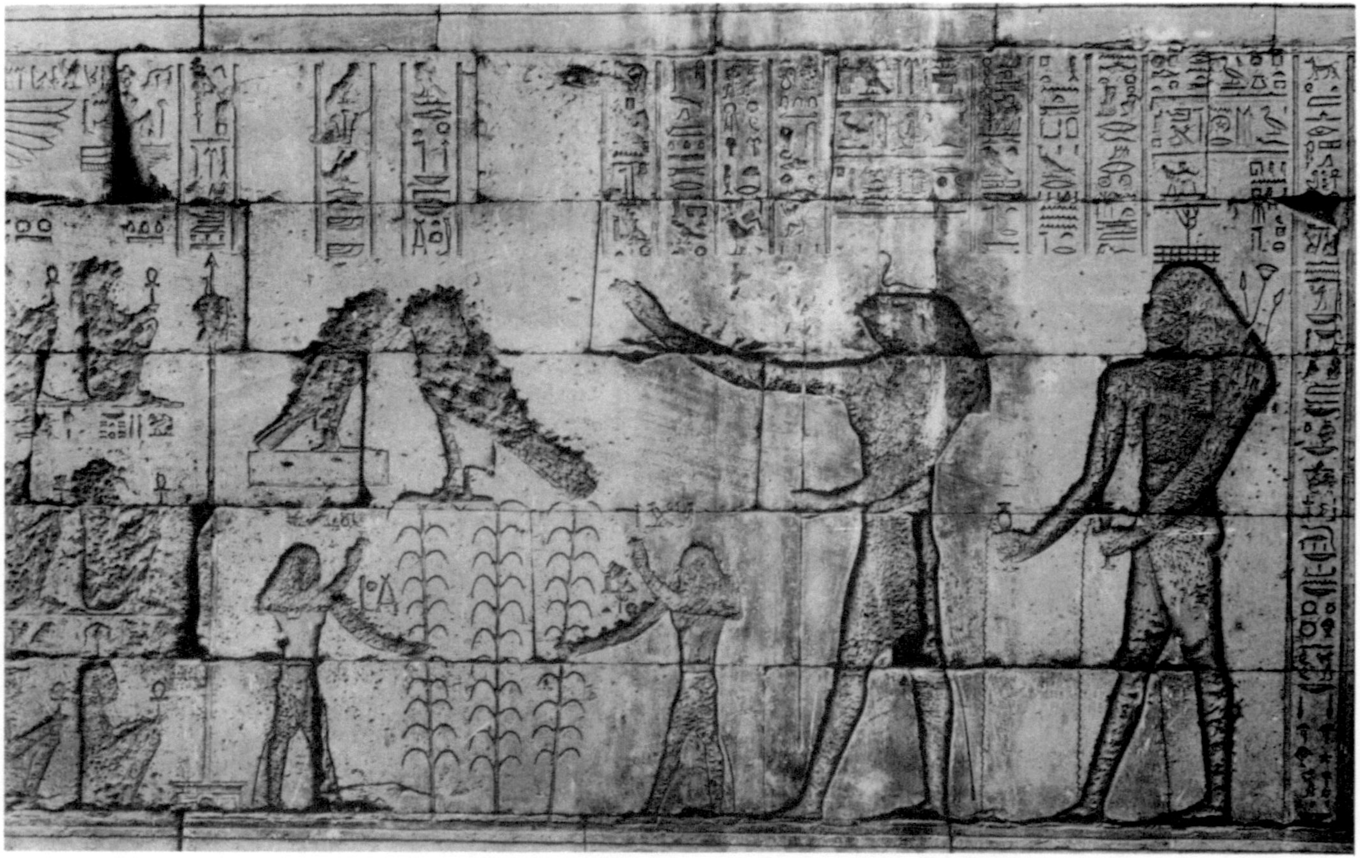

Plate 2: Reliefs to the cosmogony text on the inner face of the enclosure wall.
Centre left: Horus of *Bḥdt* as the falcon of Heaven and of *Ḏb3*, facing the Flying Ba; *W3j* and *C3* below.
Centre right: The God-of-the-temple as the Protector with his weapon in hand; behind him Hapj as *p3 Ḫnw*.
From Chassinat, *Le temple d'Edfou* XIV pl. dlxi.

Plate 3: The pylon of the temple, facing south. From Chassinat, *Le temple d'Edfou* XI pl. iii.

Plate 4: The pylon, facing north, and court. From Chassinat, *Le temple d'Edfou* IX pl. v.

Plate 5: The court, and façade of the pronaos. From Chassinat, *Le temple d'Edfou* IX pl. vi.

Plate 6: Inside the pronaos. From Chassinat, *Le temple d'Edfou* IX pl. viii.

Plate 7: View of the temple roof. From Chassinat, *Le temple d'Edfou* IX pl. x.

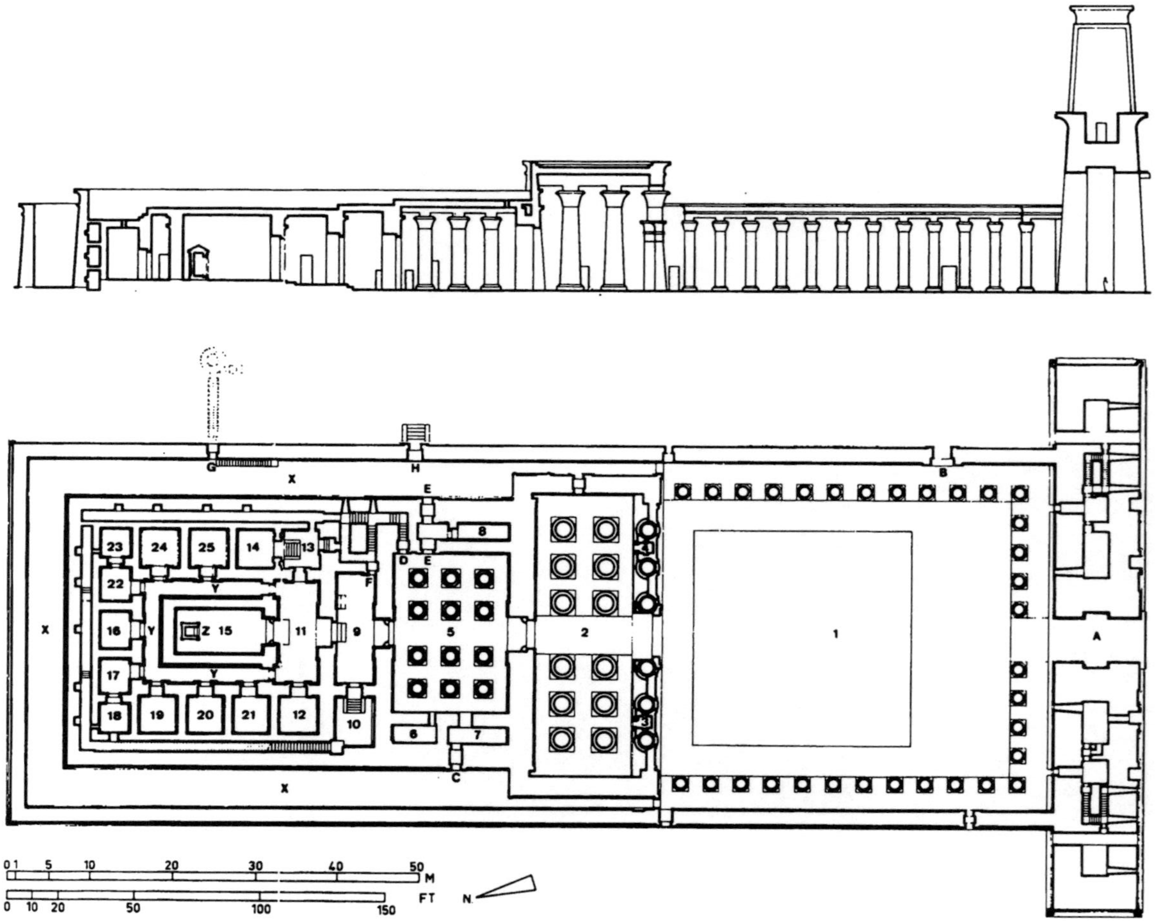

Plate 8: Ground-plan of the temple. From Sauneron and Stierlin, *Edfou et Philae.*

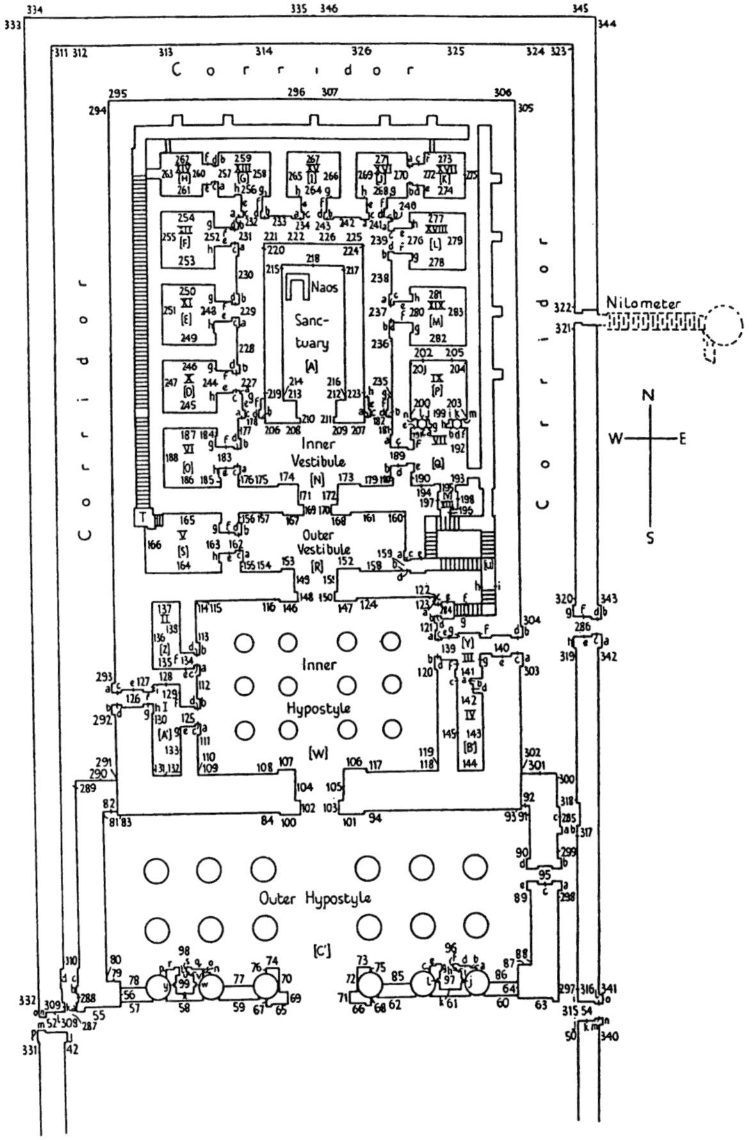

Plate 9: North part of the temple; plan. From Porter and Moss, *Topographical Bibliography of Ancient Egyptian Hieroglyphic Texts, Reliefs and Paintings* VI.